TOUGH COMPOSITE MATERIALS

TOUGH COMPOSITE MATERIALS

Recent Developments

NASA Langley Research Center
Hampton, Virginia

NOYES PUBLICATIONS
Park Ridge, New Jersey, U.S.A.

Copyright © 1985 by Noyes Publications
Library of Congress Catalog Card Number 85-16840
ISBN: 0-8155-1039-X
Printed in the United States

Published in the United States of America by
Noyes Publications
Mill Road, Park Ridge, New Jersey 07656

10 9 8 7 6 5 4 3 2 1

Library of Congress Cataloging-in-Publication Data
Main entry under title:

Tough composite materials.

　　Based on papers presented at the Tough Composite
Materials Workshop held at NASA Langley Research
Center, Hampton, Va., May 24-26, 1983 and at the
NASA/AIAA Advanced Materials Technology Seminar held
at NASA Langley Research Center, Nov. 16-17, 1982.
　　Includes bibliographies and index.
　　1. Composite materials--Congresses. 2. Transport
planes--Materials--Congresses. I. Langley Research
Center. II. Tough Composite Materials Workshop
(1983 : Hampton, Va.) III. NASA/AIAA Advanced Materials
Technology Seminar (1982 : Hampton, Va.)
TA418.9.C6T68　1985　　620.1'18　　85-16840
ISBN 0-8155-1039-X

Foreword

A series of studies on tough composite materials is presented in this book. These advanced composite materials are extremely strong, but lightweight; and they are being used as metal replacements in a variety of applications where weight reduction is important.

Consumption of advanced composites is currently increasing at a 30% rate annually, and is expected to become a $10 billion market in the late 1990s.

The material covered here provides an overview of NASA research aimed at improving composite material performance and increasing the understanding of composite material behavior. The long range goal of NASA's composites program is to exploit the full weight reduction potential of composite materials for structural applications in commercial transports. NASA sponsors and conducts a vigorous research program to accomplish this goal. New concepts, approaches, and ideas are constantly being reviewed and considered.

The book is divided into four parts which cover composite fracture toughness and impact characterization, constituent properties and interrelationships, matrix synthesis and characterization, and selected additional subjects.

This information should be of considerable value to engineers looking for tougher and stronger lightweight materials.

The information in the book is from the following documents:

Tough Composite Materials, compiled by Louis F. Vosteen, Norman J. Johnston, and Louis A. Teichman, NASA Langley Research Center, Hampton, VA 1984. The report is the proceedings of a workshop sponsored by NASA Langley Research Center, May 1983.

Advanced Materials Technology, compiled by Charles P. Blankenship and Louis A. Teichman, NASA Langley Research Center, Hampton, VA 1982. The report is the proceedings of a seminar, sponsored by NASA Langley Research Center and the American Institute of Aeronautics and Astronautics, November 1982.

The table of contents is organized in such a way as to serve as a subject index and provides easy access to the information contained in the book.

> Advanced composition and production methods developed by Noyes Publications are employed to bring this durably bound book to you in a minimum of time. Special techniques are used to close the gap between "manuscript" and "completed book." In order to keep the price of the book to a reasonable level, it has been partially reproduced by photo-offset directly from the original reports and the cost saving passed on to the reader. Due to this method of publishing, certain portions of the book may be less legible than desired.

NOTICE

> The materials in this book were prepared as accounts of work sponsored by the National Aeronautics and Space Administration and the American Institute of Aeronautics and Astronautics. Publication does not signify that the contents necessarily reflect the views and policies of the sponsoring organizations or the publisher, nor does mention of trade names or commercial products constitute endorsement or recommendation for use.

Contents and Subject Index

PART I
COMPOSITE FRACTURE TOUGHNESS
AND IMPACT CHARACTERIZATION

CHARACTERIZATION OF INTERLAMINAR CRACK GROWTH IN COMPOSITES WITH THE DOUBLE CANTILEVER BEAM SPECIMEN 2
Donald L. Hunston
    Project Objectives 2
    Cooperating Programs 3
    Approach 4
    Specimens 5
    Bulk Resin Rate and Temperature Effects 6
    Woven Composite Rate and Temperature Effects 7
    Bulk vs Composite Fracture 8
    Fiber Nesting and Bridging 9
    Data with Tough Matrices 10
    Loading Curves for Brittle and Tough Composites 11
    Matrix vs Composite Fracture Energy 12
    Future Plans 13

CHARACTERIZING DELAMINATION RESISTANCE OF TOUGHENED RESIN COMPOSITES 14
T. Kevin O'Brien
    Introduction 14
    Edge-Delamination Tension Test Measures Interlaminar Fracture
      Toughness 15
    Interlaminar Fracture Toughness of Graphite Composites 16
    Mixed-Mode Strain Energy Release Rates Determined 17
    Interlaminar Fracture Toughness of Graphite Composites
      Measured 18

Variation in Delamination Onset Strains with Layup 19
Effect of Stacking Sequence on Mixed Fracture Modes........... 20
Influence of Material Properties on G_I Percentage 21
Interlaminar Fracture Toughness of Graphite Composites 22
Delamination Failure Criteria as a Function of Mixed-Mode
  Percentage for Graphite/Epoxy Composites 23
Characterizing Delamination Resistance in Fatigue with the Edge
  Delamination Test..................................... 24
G_c as a Function of Fatigue Cycles.......................... 25
Edge Delamination Test Specimen Size Optimization............ 26
Summary.. 27
References.. 27

COMPOSITE MATERIALS CHARACTERIZATION AND DEVELOPMENT AT AFWAL................................ 28
C.E. Browning

EFFECT OF IMPACT DAMAGE AND OPEN HOLES ON THE COMPRESSION STRENGTH OF TOUGH RESIN/HIGH STRAIN FIBER LAMINATES.. 54
Jerry G. Williams

Introduction... 54
Test Specimens and Fixture................................. 55
Laminate Damage Following Impact......................... 56
Effect of Impact Energy on Damage Size..................... 57
Impact Initiated Compression Failure Modes 58
Propagation of Impact Induced Delamination.................. 59
Shear Crippling Failure Mode................................ 60
Failure of Open-Hole Specimen Loaded in Compression........... 61
Shear Crippling Initiates Open-Hole Specimen Failure 62
Impact-Damage Failure Threshold Curve 63
Tough Resin Improves Laminate Damage Tolerance.............. 64
Effect of Size of Impact Damage on Failure Strain................ 65
Open-Hole Compression Specimens.......................... 66
Stress-Strain Response for Open-Hole Specimens................ 67
Effect of Circular Holes on Compression Strength 68
Effect of High Strain Fiber on Failure Strain of Open-Hole
  Specimens 69
Open-Hole Versus Impact Strength Reduction 70
Conclusions 71
References... 71

EFFECTS OF CONSTITUENT PROPERTIES ON COMPRESSION FAILURE MECHANISMS 72
H. Thomas Hahn

Introduction... 72
Approach... 73
Matrix and Fiber Properties................................. 74

Compressive Failure Strains of Fiber Bundles 75
  A Correlation Between Compression and Tension Failure Strains 76
  Analytical Predictions 77
  Dependence of Compressive Failure Strain on Modulus Ratio 78
  Mode of Buckling Failure 79
  Buckling Failure of E-Glass Fiber Bundle 80
  Buckling Failure of Graphite Fiber Bundles 82
  SEM Micrographs of Bundles 84
  Geometric Description of Bundle Failure 86
  Conclusions 87
  References 88

THE EFFECT OF MATRIX AND FIBER PROPERTIES ON IMPACT RESISTANCE 89
  Wolf Elber
  Introduction 89
  Failure Modes in Thin Laminates (I) 90
  Failure Modes in Thin Laminates (II) 91
  Failure Modes in Thin Laminates (III) 92
  Static/Impact Comparison 93
  Thin Plate Analysis 94
  Typical Load-Displacement Relations 95
  Lower Surface Fiber Strains 96
  Plate Failure Criteria 97
  Static Penetration Tests (I) 98
  Static Penetration Tests (II) 99
  Static Penetration Tests (III) 100
  Impact Failure Criteria 101
  Energy Limits 102
  Effect of Matrix Toughness (I) 103
  Effect of Matrix Toughness (II) 104
  Effect of Matrix Toughness (III) 105
  Delamination Analysis 106
  Impact Fiber Damage 107
  Fiber Damage Analysis 108
  Summary 109
  References 110

WORKING GROUP SUMMARY: COMPOSITE FRACTURE TOUGHNESS AND IMPACT CHARACTERIZATION 111
  T.K. O'Brien

PART II
CONSTITUENT PROPERTIES AND INTERRELATIONSHIPS

FUNDAMENTAL STUDIES OF COMPOSITE TOUGHNESS 114
  Kenneth J. Bowles
  Introduction 114

Materials 115
Results 117
Neat Resin Impact Resistance 119
Load Deflection Curves 120
Composite Properties Tests 122
Composite Drop Weight Impact Test Traces 123
Concluding Remarks 124

INVESTIGATION OF TOUGHENED NEAT RESINS AND THEIR RELATIONS TO ADVANCED COMPOSITE MECHANICAL PROPERTIES 125

R.S. Zimmerman

Introduction 125
Tension Specimen Configuration 126
Torsion Specimen 127
Square Steel Mold with Dividers and Spacers Used to Cast Flat Tensile Specimens 128
Steel Mold and Rubber Funnel Used to Cast Round Dogbone Torsion Specimens 129
Silicone Rubber Mold Used to Cast the Ciba-Geigy Fibredux 914 in the Round Dogbone Shape 130
Typical Tension Failure for Flat Dogbone Specimens 131
Typical Tension Failure Surface as Seen in Scanning Electron Microscope (SEM) 132
Closeup of Tension Failure Initiation Site 133
Atypical Failed Torsion Specimen 134
Torsion Failure Surface Showing Areas of Failure 135
Closeup of Torsion Failure Initiation Site 136
Epoxy Stress-Strain Data Plotted at Six Different Temperature and Moisture Conditions for Hercules 3502, Fibredux 914, and 2220-3 .. 137
Average Values for Dry and Moisture-Saturated Tensile Strengths for Four Epoxies 139
AS4/2220-1 Graphite/Epoxy Unidirectional Composite Longitudinal Tensile Stress-Strain Response as Predicted Using Micromechanics .. 140
AS4/2220-1 Graphite/Epoxy Unidirectional Composite Transverse Stress-Strain Response as Predicted Using Micromechanics 141
AS4/2220-1 Graphite/Epoxy Unidirectional Composite Longitudinal Shear Stress-Strain Response as Predicted Using Micromechanics ... 142
AS4/2220-1 Graphite/Epoxy Unidirectional Composite, 100°C, 3.9% Moisture (ETW), 27.6 MPa (4 ksi) Transverse Tensile Applied Stress . 143
Summary 144

CONSTITUENT PROPERTY-COMPOSITE PROPERTY RELATIONSHIPS IN THERMOSET MATRICES 145

J. Diamant and R.J. Moulton

Conclusions 155
References 156

**THE EFFECT OF CROSS-LINK DENSITY ON THE TOUGHENING
MECHANISM OF ELASTOMER-MODIFIED EPOXIES**..............157
  R.A. Pearson and A.F. Yee
    Part I: Summary..157
    Part I: The Effect of Rubber Content......................158
    Part I: Tensile Dilatometry Data for Baseline Epoxy.......159
    Part I: Tensile Dilatometry Data for Elastomer-Modified Epoxy....160
    Part I: Schematic of Fracture Specimens...................161
    Part I: SEM Micrographs of Stress-Whitened Region and Fracture
      Surface of Elastomer-Modified Epoxy.....................162
    Part I: Optical Micrograph (OM) of a Fractured Elastomer-Modified
      Epoxy...163
    Part I: Optical Micrograph of Fractured Elastomer-Modified Epoxy
      Under Cross-Polarized Light.............................164
    Part I: Conclusions.......................................165
    Part II: Objectives and Approach..........................166
    Part II: List of Materials................................167
    Part II: Fracture Touchness Results.......................168
    Part II: Volume Strain Results............................169
    Part II: SEM Analysis of SEN-3PB Fracture Surfaces........171
    Part II: Optical Microscopy of Subsurface Damage..........173
    Part II: Conclusions......................................177

**FREE VOLUME CONSIDERATIONS IN THERMOPLASTIC AND
THERMOSETTING RESINS**..178
  *Robert F. Landel, A. Gupta, J. Moacanin, D. Hong, F.D. Tsay,
  S. Chen, S. Chung, R. Fedors, and M. Cigmecioglu*
    References..190

**THE CHEMICAL NATURE OF THE FIBER/RESIN INTERFACE IN
COMPOSITE MATERIALS**...191
  R. Judd Diefendorf
    Introduction..191
    Optimum Interfacial Bond Strength.........................192
    Microstructure of Carbon Fibers...........................193
    Interlaminar Shear Strength...............................194
    Surface Roughness Effects on Interfacial Bond Strength....195
    Surface Energy and Wetting of Carbon Fibers...............196
    Basal Plane and Edge Topography...........................197
    Critical Surface Energy for Wetting.......................198
    Wetting of Surfaces by Polar Liquids......................199
    Surface Energy Components.................................200
    Determination of Dispersive and Polar Surface Energy Components
      of Graphite and PTFE....................................201
    Calibrated Solid Surfaces for Epoxy Resin Studies.........202
    Surface Energy Components of Epoxy Resins and Hardness....203
    Summary of Composite Toughness............................204
    References..205

COMPOSITE PROPERTY DEPENDENCE ON THE FIBER, MATRIX, AND THE INTERPHASE 207
Lawrence T. Drzal
   Background 207
   Background–Interphase Conceptual Model 208
   Experiments 209
   Interphase: Reinforcement Surface 210
   Interphase: Reinforced Surface–Micrographs of the Single-Fiber Test 211
   Interphase: Reinforced Surface–Single-Fiber Interfacial Shear Strengths 212
   Interphase: Matrix 213
   Interphase: Matrix–Interfacial Shear Strengths 214
   Interphase: Matrix–Micrographs of the Critical-Fiber Length Test ... 215
   Interphase: Matrix–Schematic Model of the Interphase 216
   Interphase: Hygrothermal Effects 217
   Interphase: Hygrothermal Effects–125°C Exposure 218
   Interphase: Hygrothermal Effects–Fractured Surfaces Parallel to Fiber Axis 219
   Interphase: Hygrothermal Effects–Fractured Surfaces Perpendicular to Fiber Axis 220
   Summary 221
   References 222

NEWER CARBON FIBERS AND THEIR PROPERTIES 223
Roger Bacon
   Introduction 223
   Materials Development Goals 223
   Key Composite Properties 224
   Newer Carbon Fibers 225
   The Graphite Layer Plane 226
   The Graphite Crystal 227
   Fiber Structure 228
   Effects of Orientation on Properties 229
   Effects of Crystallinity on Properties 230
   Effects of Fiber Defects 231
   Carbon Fiber Manufacturing Process 232
   Control of Structure 233
   Future Carbon Fibers 234

WORKING GROUP SUMMARY: CONSTITUENT PROPERTIES AND INTERRELATIONSHIPS 235
R.F. Landel

PART III
MATRIX SYNTHESIS AND CHARACTERIZATION

DEVELOPMENT OF A HETEROGENEOUS LAMINATING RESIN 240
Rex Gosnell
   Objective 240

Polyblend Epoxy Adhesives.................................241
Mechanical Behavior of Polymeric Materials...................242
Elements of Melt Transfer Process for High-Modulus Fiber........243
Narmco Composite Impact Screening Test....................244
Impact Damage Value....................................245
Short Beam Shear..246
"All-Epoxy" Systems.....................................247
More ""All-Epoxy" Systems................................248
Bismaleimide/Epoxy Systems...............................249
Conclusions...250

MODIFIED EPOXY COMPOSITES 251
W.J. Gilwee
Conclusions...260
References..261

**MORPHOLOGY AND DYNAMIC MECHANICAL PROPERTIES OF
DIGLYCIDYL ETHER OF BISPHENOL-A TOUGHENED WITH
CARBOXYL-TERMINATED BUTADIENE-ACRYLONITRILE**.........262
*Su-Don Hong, Shirley Y. Chung, Robert F. Fedors, Jovan Moacanin
and Amitava Gupta*
Introduction...262
Compositions of Testing Specimen..........................263
Characterization of the State of Cure of Neat Resins............264
Characterization of the State of Cure of Fiber-Reinforced
 Composites...265
Morphology of Fracture Surfaces of Neat Resins................266
Morphology of Fracture Surfaces of Fiber-Reinforced Composites...267
Saxs Characterization of Particle Size and Size Distribution........268
Conclusion..269
References..270

**MATRIX RESIN CHARACTERIZATION IN CURED GRAPHITE
COMPOSITES USING DIFFUSE REFLECTANCE-FTIR**...............271
Philip R. Young and A.C. Chang
Introduction...271
Reflectance Techniques...................................272
Diffuse Reflectance Optics................................273
DR-FTIR Spectrum of Cured Graphite Composite...............274
DR-FTIR Spectrum of Celion 6000 Graphite Fiber..............275
Comparison of Spectra Obtained by Diffuse Reflectance and KBr
 Transmission..276
Comparison of Spectra Obtained by Surface Reflectance and
 Powdered Sample Techniques............................277
Comparison of Infrared Sampling Techniques..................278
Diffuse Reflectance Artifacts...............................280
Initial Applications of DR-FTIR.............................281
DR-FTIR Spectra of Graphite/Polysulfone Composite Before and
 After Radiation Exposure................................282

xiv   Contents and Subject Index

    DR-FTIR Spectra of Graphite/Epoxy Composite Before and After
       Thermal Aging 283
    DR-FTIR Spectra of Graphite/Polyimide Composite Before and
       After Thermal Aging 284
    Spectra of Thermally Aged Skybond 710 Composite and
       CAB-O-SIL... 285
    DR-FTIR Spectra of Graphite/Addition Polyimide Composite
       Before and After Thermal Aging......................... 286
    DR-FTIR Spectra of Polyimide-Sulfone Adhesive During Cure...... 287
    Comparison of Regular Press and Rapid Bonding of Polyimide-
       Sulfone Adhesive on Titanium 288
    DR-FTIR Spectra of Toughened Epoxy Composite 289
    Summary... 290
    References.. 291

SOLVENT RESISTANT THERMOPLASTIC COMPOSITE MATRICES ... 292
   P.M. Hergenrother, B.J. Jensen and S.J. Havens
    Introduction.. 292
    Modified Polysulfones with Improved Solvent Resistance 293
    Approaches Investigated................................ 294
    Synthesis of Hydroxy-Terminated Sulfone Oligomers............ 295
    Synthesis of 4-Ethynylbenzoyl Chloride 296
    Synthesis of Ethynyl-Terminated Sulfone (ETS) 297
    Isothermal Aging Performance of 250°C Cured Films 298
    Thin-Film Properties 299
    Preliminary Unidirectional Celion-6000 Laminate Properties....... 300
    Solvent Resistance (24-Hour Soak)........................ 301
    Fracture Energy of Various Polymers 302
    Sulfone/Ester Polymer Containing Pendent Ethynyl Groups 303
    Film Properties of Sulfone/Ester Polymer and UDEL............ 304
    Ethynyl-Terminated Polyarylates.......................... 305
    Phenoxy Resins Containing Pendent Ethynyl Groups............ 306
    Use of a Coreactant to Increase Crosslink Density 307
    Conclusions 308
    References.. 309

THERMOPLASTIC/MELT-PROCESSABLE POLYIMIDES............. 310
   T.L. St. Clair and H.D. Burks
    Introduction.. 310
    LARC TPI... 311
    Thermoplastic Polyimidesulfone........................... 312
    Hot-Melt Processable Polyimide 313
    Polymer Synthesis Scheme 314
    Mechanical Properties Unfilled BDSDA/APB 315
    Thermooxidative Stability Comparison 316
    Viscosity at Midrange Processing Temperature 317
    Apparent Viscosity for Processing Parameters................. 318
    Chemical Resistance.................................... 319

Encapped Processable Polyimide 320
Effect of Molecular Weight on Viscosity 321
Effect of Molecular Weight on Toughness. 322
Synthesis of Two Processable Polyimides................. 323
Viscosity/Strain Rate Data 324
Commercially Promising Processable Polyimides 325
Summary.. 326
References.. 327

ALIPHATIC-AROMATIC HETEROCYCLICS AS POTENTIAL THERMOPLASTICS FOR COMPOSITE MATRICES................ 328
Chad B. Delano and Charles J. Kiskiras
Introduction... 328
PBI-8 329
Approach.. 330
Polyimide Structures 331
Linear Aliphatic Polybenzimidazoles...................... 332
Flow by TMA... 333
Flow Curves for Aliphatic Polyimides 334
DSC Curves for PMDA/1,12-Dodecanediamine Polyimide 335
Flow Curves for Aliphatic Polyimides 336
Flow Curves for Aliphatic Polybenzimidazole.............. 337
Flow Curves for N-Arylenepolybenzimidazoles 338
Dry and Wet Polymer Properties of Polyimides............ 339
Dry and Wet Polymer Properties of Polybenzimidazoles 340
Solvent Screening of Molded Specimens................... 341
Solvent Screening of Molded Polybenzimidazoles 342
Stressed Solvent Resistance of Selected Polymers 343
Polymer Selections 344
First Year's Key Program Results........................ 345

WORKING GROUP SUMMARY: MATRIX SYNTHESIS AND CHARACTERIZATION..................................... 346
P.M. Hergenrother

PART IV
SELECTED ADDITIONAL SUBJECTS

POLYMER MATRIX COMPOSITES RESEARCH AT NASA LEWIS RESEARCH CENTER 350
T.T. Serafini
Current Program Thrusts............................... 350
Lewis PMR Polyimide Technology........................ 351
Monomers Used for PMR-15 Polyimide 352
Versatility of PMR Approach............................ 353
PMR-15 Polyimide Modifications for Improved Prepreg Tack....... 354
Lower-Curing-Temperature PMR Polyimides................ 355
Improved Celion 6000/PMR-15 Composites 356

Improved Shear Strain of Imide-Modified Epoxy 357
Composite Durability 358
Applications of PMR-15 Polyimide Composites 359
Ultra-High-Tip-Speed Fan Blades 360
Application of Composites on Quiet Clean Short-Haul Experimental
Engine (QCSEE) 361
Graphite Fiber/PMR-15 QCSEE Inner Cowl 362
General Electric F404-GE-400 Turbofan 363
Graphite Fiber/PMR-15 Polyimide Outer Duct for GE F404 Engine . . 364
Graphite Fiber/PMR-15 Polyimide Outer Duct on GE F404 Engine . . 365
Schematic of GE T700 PMR-15 Composite Swirl Frame 366
Applications of Graphite Fiber/PMR-15 Polyimide Composites on
PW1120 Engine 367
DC-9 Drag Reduction 368
Kevlar Fabric/PMR-15 Reverser Stang Fairing 369
Glass Fabric/PMR-15 Beam Shield Installed on Mercury Ion
Thruster 370
Concluding Remarks 371
References 372

**FATIGUE AND FRACTURE RESEARCH IN COMPOSITE
MATERIALS** 374
T. Kevin O'Brien
Introduction 374
A Unifying Strain Criterion to Predict Fracture Toughness of
Composite Laminates 375
Measured and Predicted Values of Fracture Toughness 376
Improving Damage Tolerance of Composite Laminates with Buffer
Strips—Test Results 377
Improving Damage Tolerance of Composite Laminates with Buffer
Strips—Analysis 378
Hole Elongation Under Cyclic Loading in Bolted Composite Joints . . . 379
Bolt Clampup Relaxation 380
3-D Stresses at Laminate Bolt Holes 381
Damage Threshold Under Low-Velocity Impact of Composites 382
Composite Impact Screening by Static Indentation Tests 383
Strain Energy Release Rate for Delamination Growth Determined . . . 384
Delamination Onset Predicted 385
Edge Delamination Tension Test Measures Interlaminar Fracture
Toughness 386
Mixed-Mode Strain Energy Release Rates Determined 387
Interlaminar Fracture Toughness of Graphite Composites Measured . . 388
Investigation of Instability-Related Delamination Growth 389
Local Delamination Causes Tensile Fatigue Failures 390
Test Methods for Compression Fatigue of Composites 391
Fatigue Damage in Boron/Aluminum Laminates 392
Debonding of Adhesively Bonded Composites Under Fatigue
Loading 393

Future Work. 394
References 395

PROCESSING COMPOSITE MATERIALS. 397
R.M. Baucom
Introduction. 397
Composite Processing and Applications 398
Composite Molding Methods 399
Composite Molding by Thermal Expansion. 400
Autoclave Molding of Composites. 401
Typical Composite Cure Cycle 402
Ultrasonic Inspection of Composites 403
Inspection Techniques 404
DC-10 Upper Aft Rudders. 405
Composite DC-10 Upper Aft Rudder Advantages. 406
DC-10 Upper Aft Rudder Manufacturing Sequence 407
Thermal Expansion Molding Technique. 408
L-1011 Advanced Composite Vertical Fin 409
L-1011 ACVF Structural Configuration. 410
ACVF Skin Lay-Up Schematic. 411
ACVF Hat/Skin Vacuum Bag Assembly. 412
ACVF Cover Fabrication 413
Complete Graphite/Epoxy ACVF Cover 414
Automated Integrated Manufacturing System. 415
Integrated Laminating Center 416
Hot Melt Fusion Composites 417
Kevlar/Epoxy Faceguard. 418
Composite Wheelchair Project 419
Summary. 420

**OPPORTUNITIES FOR COMPOSITES IN COMMERCIAL TRANSPORT
STRUCTURES** 421
Herman L. Bohon
ACEE Composites Program—Introduction 421
Component Development Toward a Production Commitment 422
ACEE Composite Secondary Structures. 423
DC-10 Graphite/Epoxy Upper Aft Rudder 424
727 Graphite/Epoxy Elevator 425
L-1011 Graphite/Epoxy Aileron. 426
ACEE Composite Medium Primary Structures. 427
DC-10 Graphite/Epoxy Vertical Stabilizer 428
737 Graphite/Epoxy Horizontal Stabilizer 429
L-1011 Graphite/Epoxy Vertical Fin. 430
ACEE Composite Component Status. 431
Characteristics of Graphite/Epoxy Material 432
Typical Cure Cycles for Graphite/Epoxy 433
Large Complex Structures Manufactured 434
Lightweight Structural Components Simplify Final Assembly 435

Structural Performance Demonstrated. 436
Spar Static Strength Tests for Manufacturing Variances 437
Comparison of Static Strength for Structural Materials. 438
Influence of Environmental Parameters on Structural Performance. . . 439
Influence of Moisture and Temperature on Strain 440
Key Areas for Technology Advancement 441
Effects of Secondary Loads in Composites. 442
Influence of Load Cycling on Interlamina Strength 443
Influence of Damage on Allowable Strain. 444
Weight Efficiency of Primary Structure Dependent on Allowable
  Strain 445
ACEE Composites Program Conclusions 446

DURABILITY OF AIRCRAFT COMPOSITE MATERIALS 447
H. Benson Dexter

Introduction. 447
Flight Service Composite Components on Transport Aircraft. 448
ACEE Composite Secondary Structures. 449
Flight Service Composite Components on Helicopters 450
Bell 206L Helicopter Composite Components. 451
Sikorsky S-76 Helicopter Composite Components 452
NASA Composite Structures Flight Service Summary. 453
Residual Strength of Graphite/Epoxy Spoilers 454
NASA Composite Component Inspection and Maintenance Results. . . 455
B-737 Spoiler In-Service Damage and Repair. 456
DC-10 Composite Rudder Lightning Damage 457
Environmental Effects on Long-Term Durability of Composite
  Materials for Commercial Aircraft. 458
Worldwide Environmental Exposure of Composite Materials 459
Moisture Absorption During Ground Exposure 460
Residual Strength of Composite Materials After Worldwide
  Outdoor Exposure. 461
Residual Tensile Strength After Sustained Stress Outdoor
  Exposures 462
Effect of Aircraft Fluids on Composite Materials After 5 Years
  of Exposure 463
Surface Degradation of AS/3501 Graphite/Epoxy 464
Concluding Remarks 465
References. 466

Part I

Composite Fracture Toughness and Impact Characterization

The information in Part I is from *Tough Composite Materials,* compiled by Louis F. Vosteen, Norman J. Johnston, and Louis A. Teichman, NASA Langley Research Center, Hampton, VA, 1984. The report is the proceedings of a workshop sponsored by NASA Langley Research Center, May 1983.

CHARACTERIZATION OF INTERLAMINAR CRACK GROWTH IN COMPOSITES WITH THE DOUBLE CANTILEVER BEAM SPECIMEN

Donald L. Hunston

National Bureau of Standards
Polymers Division
Washington, D.C.

PROJECT OBJECTIVES

Since delamination is considered to be a major failure mechanism in composites, there is a need and desire to develop test methods that assess the resistance of various composite materials to the growth of interlaminar cracks. In response to this need our program is examining the double cantilever beam (DCB) specimen. The program has two objectives (Figure 1). The first is to explore the DCB as a quantitative test method. This involves both a review of the results of other workers in this field and the completion of experiments in areas where additional work is required. The second objective is to investigate the micromechanics of failure for composites with tough matrix resins from certain generic types of polymeric systems: brittle thermosets, toughened thermosets, and tough thermoplastics. From this information it is hoped that a better picture of toughening in composites can be formulated.

EXPLORE DCB SPECIMEN AS A QUANTITATIVE TEST METHOD

INVESTIGATE MICROMECHANICS OF FAILURE WITH TOUGH MATRIX RESINS

Figure 1

COOPERATING PROGRAMS

The project is part of a larger effort involving a number of groups (Figure 2). Our work which focuses on the DCB specimen and an examination of failure mechanisms is coordinated with programs at NASA studying the edge delamination specimen and developing specialized materials. Through this combination, data on a variety of loading geometries and materials can be obtained. To assist in the analyses of the test methods, a cooperative project under the direction of Prof. S. S. Wang at the University of Illinois is developing stress analyses for the various specimen geometries using realistic constitutive equations. Finally, there is close coordination with industry including Bill Bascom at Hercules and Rich Moulton at Hexcel who are supplying commercial and model materials for study.

DELAMINATION PROGRAM:

NBS: DCB
FAILURE MECHANICS

NASA: EDGE DELAMINATION
SPECIALIZED MATERIALS

UNIV. OF ILL.: STRESS ANALYSIS

INDUSTRIES: COMMERCIAL & MODEL MATERIALS

Figure 2

APPROACH

The approach in this program is to examine in a systematic way all of the variables that could affect the DCB test. These variables, listed in Figure 3, are relatively obvious although the inclusion of analysis method deserves some comment. A number of different procedures are being used by various groups to evaluate a fracture energy from their DCB results. In the ideal case of a linear elastic material the analysis method should make no difference. With the tougher matrix resins, however, some deviations from ideal behavior can occur and thus the analysis method may make a difference. Consequently, this is being examined. The variables listed in Figure 3 are divided into 4 groups to indicate the approximate order in which they are being addressed. There are however strong interactions between these parameters and thus they cannot be entirely separated. The recent work on the effects of specimen geometry and lay-up will be covered in forthcoming publications. The purpose of this paper will be to focus on preliminary results in two other areas: the effects of temperature and loading rate for woven composites, and the effects of matrix toughening in woven and unidirectional composites.

VARIABLES

	Lay-up	(I)
	Dimensions & Shape	(II)
●	Temperature & Loading Rate	(II)
●	Matrix Material & Fiber Type	(III)
	Fiber Volume Fraction	(III)
	Environment (Humidity)	(IV)

ANALYSIS METHOD (I)

Compliance Measurement

Area Measurement

Compliance Measurement with Non-Linear Bending

Tapered Beam with Bending Modulus

Other

Figure 3

SPECIMENS

The tests are examining the failure behavior of both the composite and bulk samples of the various resins. The bulk (or neat) tests are conducted with standard compact tension specimens. The composite specimens under examination are the simple double cantilever beam (DCB) and the width tapered DCB (see Figure 4) with various degrees of taper. The untapered DCB may have an advantage in being more easily addressed in a stress analysis. The tapered DCB may have a potential advantage in that the crack grows at a constant load thus simplifying the data analysis. Both woven reinforcement and unidirectional specimens are being fabricated with the 0 fiber direction defined to be down the length of the specimen. A wide range of specimens are being examined but the results reported here involve 12 ply woven reinforcement composites and 0_{24} unidirectional composites.

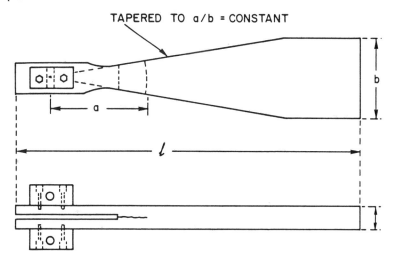

Figure 4

BULK RESIN RATE AND TEMPERATURE EFFECTS

The first topic to be examined here is the effects of temperature and loading rate (i.e., cross head speed of test machine). The motivation for the interest in these effects can be illustrated in Figure 5 which shows the fracture data for bulk samples of a model rubber-modified epoxy. With an unmodified epoxy the fracture energy for both bulk and composite specimens shows relatively small variations with moderate changes in temperature and loading rate. For tough materials, like the modified epoxy, however, the sensitivity of bulk samples to these variables is very large. In Figure 5, for example, the fracture energies vary by more than an order of magnitude depending on test conditions. In light of this behavior, there is a need to examine the effects of these parameters on the fracture properties of composites made with tough resins.

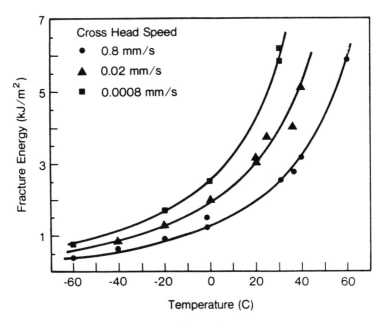

Figure 5

WOVEN COMPOSITE RATE AND TEMPERATURE EFFECTS

It has been shown that improving the fracture energy of the resin itself by 25 fold will increase the interlaminar fracture energy as measured by the DCB of a woven composite by 4 to 8 fold. This difference has been explained by noting that the high toughness of the resin is associated with a large crack tip deformation zone. In the composite, however, the deformation zone is restricted by the presence of the fibers. Consequently, there may be a potential for deformation in the matrix that is not utilized. If the temperature is lowered or the loading rate is increased, perhaps this reserve can be called upon. Such a hypothesis might suggest that composites fabricated with a tough resin might show less dependence on rate and temperature than the resin itself. Figure 6 gives some preliminary results for woven reinforcement composites made with a rubber-modified epoxy (2 fold improvement in composite fracture energy relative to the unmodified epoxy). The results show relatively little rate and temperature dependence over the range of conditions tested. Additional studies are now under way to examine these trends in more detail, particularly with unidirectional composites.

Temperature	Fracture Energy (kJ/m^2) at indicated Cross-Head Speed (mm/sec)		
°C	0.0008	0.04	0.8
40	1.7	1.8	1.8
24	1.7	1.7	1.9
0	1.7	1.9	2.0
-25	1.9	2.1	2.4

Figure 6

BULK VS COMPOSITE FRACTURE

The second area of interest in this paper is the effects of varying the matrix resin toughness on the interlaminar fracture energy. Figure 7 shows baseline data on bulk specimens and woven and unidirectional composites all made with a model epoxy. The two unidirectional composites differ only in that one was fabricated by Hexcel and the other by NASA Langley. Note first the relative values for the bulk samples and the Hexcel composite. The interlaminar fracture energy is slightly higher than that for the resin indicating a rough transfer from the resin to the composite with some enhancement of toughness in the composite through additional mechanisms. Both fiber breakage and crack bifurcation were observed in the composite fracture surfaces. The woven composite exhibits significant data scatter but the fracture energy is clearly higher than that of the resin or the Hexcel composite. The crack propagation in the woven composite was stick slip with initiation and arrest at the crossover points in the weave. Consequently, one possible explanation for the enhanced toughness may be the addition of crack pinning as an added toughening mechanism. Compare now the NASA and Hexcel composites. The NASA samples clearly give a higher fracture energy.

ELASTOMER CONCENTRATION	FRACTURE ENERGY (kJ/m^2)			
	NEAT	WOVEN	0_{24} (HEXCEL)	0_{24} (NASA)
0%	0.23	$0.6-1.3^A$	0.37	0.77

^{A}NUMBERS INDICATE RANGE OF VALUES FOR 10 SPECIMENS.

Figure 7

Composite Fracture Toughness and Impact Characterization   9

FIBER NESTING AND BRIDGING

This comparison illustrates the importance of factors other than matrix type in determining interlaminar fracture behavior. The NASA and Hexcel samples were not fabricated with the same procedure and this produced differences in the fiber distribution. As a result the NASA materials exhibited substantially more nesting and fiber bridging. In nesting, the fibers from different plies mix together so the crack can not follow a planar surface between plies. If viewed from the end of the specimen, the crack front is not linear because it must go around the interpenetrating fibers (Figure 8). Fiber bridging is a related effect. When nesting is present, the crack front may not completely separate all the fibers on one side of the crack from those on the other. These unseparated fibers can then span across the crack behind the crack tip as shown in Figure 8. From the results in this work and other similar studies elsewhere, it appears that excessive fiber nesting and bridging in a specimen will cause an elevation in the measured interlaminar fracture energy.

Nesting   Fiber Bridging

Figure 8

DATA WITH TOUGH MATRICES

Turn now to the effect of toughening the matrix resin. Figure 9 confirms the previous results, which found that an improvement of 25 fold in the resin fracture energy gives a 4 to 8 fold improvement in woven composite results. With unidirectional composites the improvement was expected to be much less because there are no resin rich areas equivalent to those at the cross-over points in the weave in woven composites. Surprisingly, however, the improvement in the unidirectional specimens, 2.5 to 5 fold, was almost as large as that found for woven composites. Moreover, the additional toughening mechanisms (nesting, fiber bridging, crack pinning, etc.) that were suggested in the unmodified epoxy results appear to be present in the toughened composite as well but perhaps to a lesser degree.

ELASTOMER CONCENTRATION	FRACTURE ENERGY (kJ/m^2)			
	NEAT	WOVEN	0_{24} (HEXCEL)	0_{24} (NASA)
0%	0.23	0.6-1.3[A]	0.37	0.77
13.5%	5.8	3.6-4.6[A]	1.6-2.0[B]	1.9-2.3[B]

[A] NUMBERS INDICATE RANGE OF VALUES FOR 10 SPECIMENS.

[B] LOWER NUMBER CORRESPONDS TO FIRST INDICATION OF CRACK GROWTH, HIGHER NUMBER STEADY CRACK GROWTH AT 0.02 CM/S.

Figure 9

LOADING CURVES FOR BRITTLE AND TOUGH COMPOSITES

In the unidirectional composites the DCB test measures the strain energy release rate corresponding to relatively steady crack growth. With the brittle materials the first indication of crack growth in the composite occurs at about the same strain energy release rate as the steady growth. Figure 10 shows an example of the loading curves for such a composite with the tapered DCB specimen. When tough matrix resins are used, the first indication of crack growth occurs at a somewhat lower strain energy release rate than that corresponding to steady growth. This behavior can be seen either by observing the crack or by noting deviation from linearity in the load-displacement curves (Figure 10). As a result, for the tougher matrix resin composites, two fracture energies are given: an onset and a steady growth value. The onset number however may be somewhat subjective since its exact value may depend on how closely the data are examined. Nevertheless, the difference between the onset and steady growth values provides a measure of the magnitude of the effect.

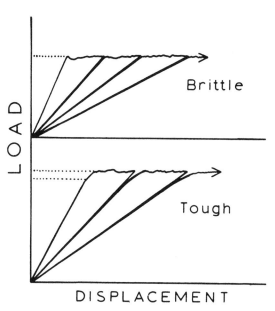

Figure 10

MATRIX VS COMPOSITE FRACTURE ENERGY

The results in Figure 9 show that a 25 fold improvement in resin fracture energy gives only a 4 fold improvement in composite interlaminar fracture energy. Since toughening of the resin usually degrades other desirable properties, it is of interest to see if a 10 fold improvement in resin fracture energy would still produce a 4 fold improvement in the composite. Figure 11 shows the results for the Hexcel materials in Figure 9 plus an intermediate toughness resin (8% rubber) also fabricated by Hexcel ($\square$). The results clearly indicate that an intermediate toughness resin produces and intermediate toughness composite. These limited data suggest that each 4 J/m^2 increase in resin fracture energy gives about 1 J/m^2 improvement in interlaminar fracture energy. Thus there is a trade-off between interlaminar toughness and other properties. Also shown on this graph are results for polysulfone ($\triangle$) and polyetherimide ($\diamondsuit$). Although the results are preliminary, these materials seem to follow the same general trend as the modified epoxies. The fracture surface of the polyetherimide exhibited significantly more interfacial failure than the other samples and thus inferior bonding may explain why this composite appears to fall somewhat below the general trend.

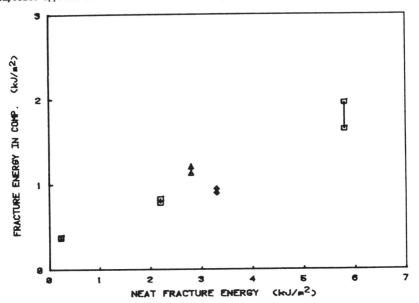

Figure 11

FUTURE PLANS

Future plans in this program (Figure 12) involve completing the study of phase I and II parameters. In addition the effects of matrix toughness will be included since a tough matrix may affect the results of experiments examining these variables. For example, preliminary experiments have suggested that with tougher matrix resins it is desirable to have stiffer specimens to minimize the effects of non-linear bending and deviation from linear elastic behavior in the resin. With regard to rate and temperature effects, the major focus will be on unidirectional composites since the behavior shown in Figure 10 suggests these variables may be important in this case. Thermoplastic matrix resins are of particular interest here since even the behavior in bulk has not been studied in detail. Finally, the implications of these studies in terms of the analysis method will be given attention because the test can only be as good as the approximations and assumptions involved in the fracture energy calculation, and with tough matrices more uncertainty is now present in this regard.

VARIABLES

LAY-UP	(I)
DIMENSIONS & SHAPE	(II)
TEMPERATURE & LOADING RATE	(II)
MATRIX MATERIAL & FIBER TYPE	(III)
FIBER VOLUME FRACTION	(III)
ENVIRONMENT (HUMIDITY)	(IV)

ANALYSIS METHOD   (I)

COMPLIANCE MEASUREMENT

AREA MEASUREMENT

COMPLIANCE MEASUREMENT WITH NON-LINEAR
   BENDING

TAPERED BEAM WITH BENDING MODULUS

OTHER

Figure 12

CHARACTERIZING DELAMINATION RESISTANCE OF TOUGHENED RESIN COMPOSITES

T. Kevin O'Brien

Structures Laboratory
U.S. Army Research and Technology Laboratories (AVRADCOM)
NASA Langley Research Center
Hampton, Virginia

INTRODUCTION

One major obstacle to the efficient application of advanced composite materials in large primary aircraft structures is the tendency for these materials to delaminate. Because delamination may adversely influence stiffness, strength, and fatigue life, composites with toughened matrix resins are currently being developed to improve delamination resistance. However, whereas the fracture toughness of new resins can be measured in the bulk, no sound relationships yet exist to relate bulk fracture to delamination resistance in the composite. Therefore, investigators have measured the interlaminar fracture toughness of the composite in order to screen new toughened matrix resins for improved delamination resistance.

Interlaminar fracture can occur under a mixture of mode I (interlaminar tension or peel) and mode II (interlaminar shear). These modes are usually mixed in practice; hence the term "mixed-mode fracture." Of the specimens used to measure interlaminar fracture toughness, the double-cantilever-beam specimen delaminates under mode I only, whereas the edge delamination test specimen delaminates under a mixed mode. The purpose of this investigation was to determine the influence of mixed-mode fracture on interlaminar fracture toughness for "brittle" and "tough" matrices subjected to both static and cyclic loads, using the edge delamination test.

EDGE-DELAMINATION TENSION TEST MEASURES INTERLAMINAR FRACTURE TOUGHNESS

A simple test has been developed for measuring the interlaminar fracture toughness of composites made with toughened matrix resins (ref. 1). The test involves measuring the stiffness, E_{LAM}, and nominal strain at onset of delamination, ε_c, during a tension test of an 11-ply $[\pm30/\pm30/90/\overline{90}]_s$ laminate (fig. 1). These quantities, along with the measured thickness t, are substituted into a closed-form equation for the strain energy release rate, G, for edge delamination growth in an unnotched laminate (ref. 2). The E* term in the equation is the stiffness of the $[\pm30/\pm30/90/\overline{90}]_s$ laminate if the 30/90 interfaces were completely delaminated. It can be calculated from the simple rule of mixtures equation shown in figure 1 by substituting the laminate stiffness measured during tension tests of $[\pm30]_s$ and $[90]_n$ laminates. The critical value of G_c at delamination onset is a measure of the interlaminar fracture toughness of the composite. This edge delamination test is being used by Boeing, Douglas, and Lockheed under the NASA ACEE (Aircraft Energy Efficiency) key technologies contracts to screen toughened resin composites for improved delamination resistance (ref. 3).

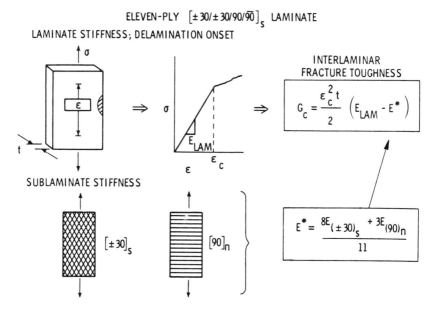

Figure 1

INTERLAMINAR FRACTURE TOUGHNESS OF GRAPHITE COMPOSITES

Figure 2 illustrates how the edge delamination test (EDT) can rank the relative delamination resistance of graphite composites with different matrix resins. Results are shown for four matrices, ranging from a very brittle 350°F cure epoxy to a tougher thermoplastic, and from 250°F cure epoxy to a very tough "rubber-toughened" epoxy. Also shown in figure 2 are double-cantilever-beam (DCB) measurements for the same composites. Results indicate that either test will yield a qualitative ranking of improvements in delamination resistance. However, the G_{Ic} and G_c values differ because the DCB test involves only interlaminar tension, whereas the edge delamination test involves a combination of interlaminar tension and shear.

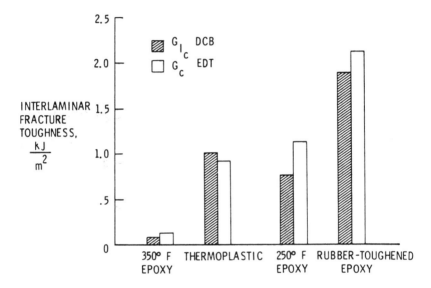

Figure 2

MIXED-MODE STRAIN ENERGY RELEASE RATES DETERMINED

A quasi-three-dimensional finite-element analysis (ref. 4) was performed to determine the relative crack-opening (mode I) and shear (mode II) contributions of the $[\pm 30/\pm 30/90/\overline{90}]_s$ edge delamination specimen (ref. 1). Delaminations were modeled in the -30/90 interfaces where they were observed to occur in experiments. Figure 3 indicates that the total G represented by G_I plus G_{II} reaches a value prescribed by the closed-form equation derived from laminated-plate theory and the rule of mixtures. Furthermore, like the total G, the G_I and G_{II} components are also independent of delamination size.

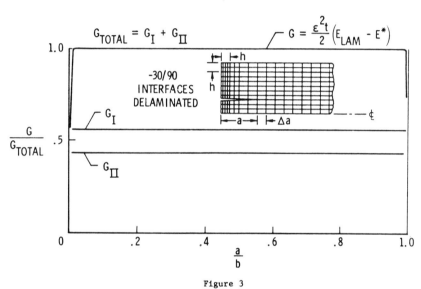

Figure 3

INTERLAMINAR FRACTURE TOUGHNESS OF GRAPHITE COMPOSITES MEASURED

Figure 4 shows results (ref. 1) of pure crack-opening (mode I) double-cantilever-beam tests and edge delamination tension tests for a relatively brittle 350°F cure epoxy (5208), a tougher 250°F cure epoxy (H205), and a still tougher rubber-toughened 250°F cure epoxy (F185). Results indicate that for the brittle epoxy, even in the mixed-mode test, only the crack-opening fracture mode contributes to delamination. However, for the tougher 250°F cure epoxy and its rubber-toughened version, both the crack-opening and shear fracture modes contribute to delamination. Hence, although both tests indicate relative improvements among materials, one test alone is not sufficient to quantify interlaminar fracture toughness.

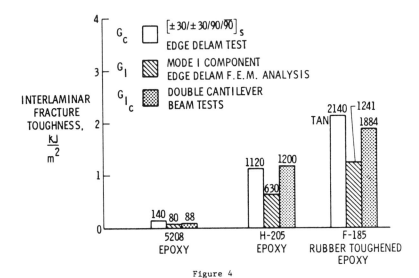

Figure 4

VARIATION IN DELAMINATION ONSET STRAINS WITH LAYUP

A parametric study was conducted (ref. 5) to optimize layups for the edge delamination test. Two families of layups were considered. The first, $[\pm\theta/\overline{90}]_s$, contained only angle plies and 90° plies. The second, $[\pm\theta/0/\overline{90}]_s$, contained angle plies, 0° plies, and 90° plies. Figure 5 shows the critical strain at delamination onset, ε_c, assuming a G_c of 0.15 kJ/m², required to create an edge delamination in the two layup families. As shown in figure 5, the lowest ε_c occurs for θ in the vicinity of 30° and 35° for the $[\pm\theta/\overline{90}]_s$ and $[\pm\theta/0/\overline{90}]_s$ families, respectively. Hence, the $[\pm30/\overline{90}]_s$ family and the $[\pm35/0/\overline{90}]_s$ families appear to be good candidates for the edge delamination test.

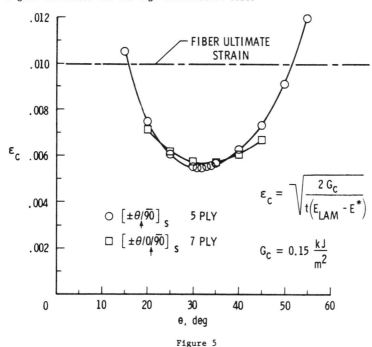

Figure 5

EFFECT OF STACKING SEQUENCE ON MIXED FRACTURE MODES

A finite-element analysis was performed in reference 5 to determine the crack-opening mode, G_I, and interlaminar shear mode, G_{II}, components of the total strain energy release rate, G, for the optimized layups of the edge delamination test. Three permutations of the $[\pm 35/0/90]_s$ layup family were analyzed. They were $[\pm 35/0/90]_s$, $[+35/0/-35/90]_s$, and $[0/\pm 35/90]_s$. All three layups had the same total G for a given nominal strain ε, but each had very different percentages of G_I and G_{II}. Figure 6 shows results for the $[\pm 35/0/90]_s$ family, as well as results for a similar quasi-isotropic family and for the original $[\pm 30/\pm 30/90/\overline{90}]_s$ layup. In all cases the delamination is modeled between the 90° ply and the adjacent plies. For the $[\pm \theta/0/90]_s$ layups, the G_I percentages range from very high to intermediate to very low when the 0° plies are either next to the interior 90° ply, between the $+\theta°$ and $-\theta°$ plies, or on the outside, respectively. Hence, the three permutations of the $[\pm 35/0/90]_s$ layup may be useful for evaluating the fracture mode dependence of composites with different matrices.

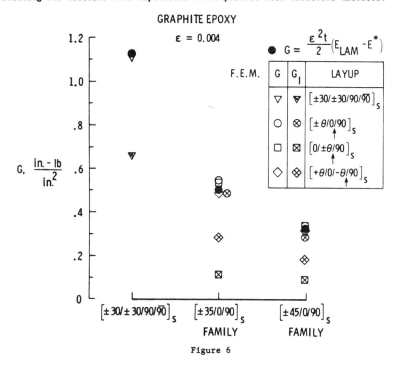

Figure 6

INFLUENCE OF MATERIAL PROPERTIES ON G_I PERCENTAGE

A parametric study was performed, using finite-element analyses, to determine the influence of material properties on the G_I and G_{II} components for all of the EDT layups. The results are shown in figure 7 for the $[\pm 35/0/90]_s$, $[+35/0/-35/90]_s$, $[0/\pm 35/90]_s$, and $[\pm 30/\pm 30/90/\overline{90}]_s$ layups made with several materials that have widely differing lamina properties. The G_I percentages are essentially constant for a given layup, independent of material properties. This material independence eliminates the need to run finite-element analyses for each new material evaluated. Hence, tests conducted on high G_I $[\pm 35/0/90]_s$, intermediate G_I $[\pm 30/\pm 30/90/\overline{90}]_s$ or $[+35/0/-35/90]_s$, and low G_I $[0/\pm 35/90]_s$ laminates may be used to quantify the interlaminar fracture mode dependence of various composite materials.

	LAMINA PROPERTIES OF GRAPHITE COMPOSITES				G_I PERCENTAGE			
	E_{11}, GPa	E_{22}, GPa	G_{12}, GPa	ν_{12}	$[\pm 35/0/90]_s$	$[\pm 35/0-35/90]_s$	$[0/\pm 35/90]_s$	$[\pm 30/\pm 30/90/\overline{90}]_s$
T300/5208	134	10.2	5.5	0.30	90%	58%	22%	57%
C6000/H205	124	8.4	5.3	0.33	88%	59%	25%	57%
C6000/F185	119	6.2	2.6	0.34	90%	61%	23%	58%

Figure 7

INTERLAMINAR FRACTURE TOUGHNESS OF GRAPHITE COMPOSITES

Figure 8 shows mean values of G_c and their G_I components measured using three EDT layups (ref. 6) for graphite composites reinforced with brittle (5208 and 2220-1) and tough (H205 and F185) resin matrices. Also shown in figure 8 are G_{Ic} values measured from DCB tests on graphite composites made with the same resin matrices. All four tests (DCB and the three EDT layups) show improvements in interlaminar fracture toughness for the tougher resin composites (H205 and F185) compared to the brittle resin composites (5208 and 2220-1). Delamination onset occurred in the mixed-mode EDT test of the 5208 composite when the G_I component reached G_{Ic} as measured by the DCB test. Furthermore, although G_c measurements of the two 2220-1 EDT layups differed, delamination occurred at identical low values of G_I. Therefore, it appears that only the interlaminar tension, G_I, fracture mode contributes to delamination in brittle resin composites subjected to mixed-mode loading. However, this was not true for the tougher H205 and F185 specimens, where interlaminar shear did contribute to delamination. For the three EDT layups with widely different G_I percentages, delamination occurred when G_I was below G_{Ic} as measured by DCB tests (fig. 8). However, the total G_c mean values for each layup were different. The apparent G_c was increasing with decreasing G_I percentage. Hence, the interlaminar shear was only partially responsible for delamination.

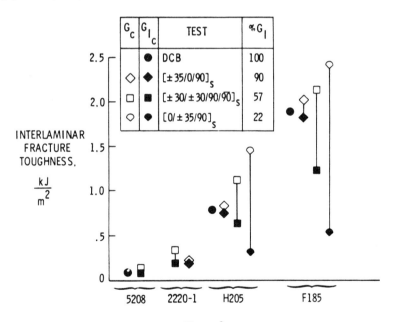

Figure 8

DELAMINATION FAILURE CRITERIA AS A FUNCTION OF MIXED-MODE PERCENTAGE FOR GRAPHITE/EPOXY COMPOSITES

The observations in figure 8 demonstrated the need for delamination failure criteria that would reflect the observed fracture mode dependence. Figure 9 shows a plot of average interlaminar fracture toughness as a function of G_I percentage for the four materials tested. For the brittle resin matrix composites (5208 and 2220-1), the interlaminar fracture toughness is simply a single value of G_{Ic}, independent of the G_I percentage. However, for the tougher resin matrix composites (H205 and F185) the interlaminar shear also contributes to delamination; therefore, a more appropriate failure criterion for these materials is a linear decrease in G_c with increasing G_I percentage. Hence, if the G_I percentage of a particular delamination failure in a structure made from these materials can be determined, then the appropriate G_c may be chosen from figure 9 to predict delamination extension.

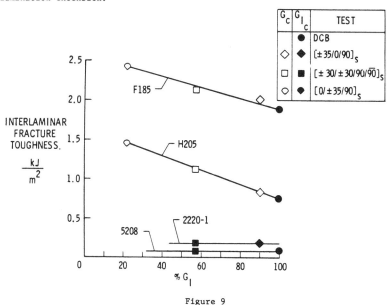

Figure 9

CHARACTERIZING DELAMINATION RESISTANCE IN FATIGUE WITH THE EDGE DELAMINATION TEST

Cyclic loading may cause extensive delamination in graphite composites, even for laminates that do not delaminate under static loads. Therefore, it is necessary to characterize delamination resistance in fatigue as well as in static loading. To this end, various EDT layups made with different resin matrices were cycled to strain levels below the delamination onset strain measured in the static tests (fig. 10). Delaminations formed at these lower cyclic strains after a certain number of cycles, N. Strain energy release rate (G) values, calculated from maximum cyclic strains and plotted as a function of the number of cycles to delamination onset, dropped sharply and then reached a plateau tantamount to a threshold for delamination onset in fatigue.

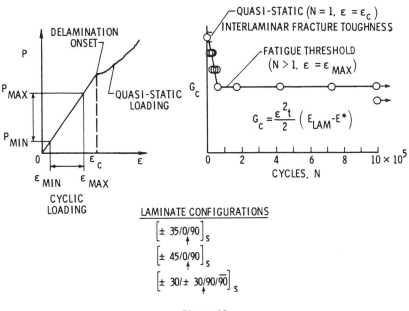

Figure 10

G_c AS A FUNCTION OF FATIGUE CYCLES

Figure 11 summarizes results of static and fatigue tests for $[\pm45/0/90]_s$ T300/5208 laminates and for $[\pm35/0/90]_s$ and $[0/\pm35/90]_s$ C6000/H205 laminates using the technique outlined in figure 10. The total G is identical for the $[\pm35/0/90]_s$ and $[0/\pm35/90]_s$ laminates, but the G_I percentages are different, as shown in figure 7. For static tests, the G_c for these two layups is different; however, the G_c thresholds under cyclic loading for the two layups are nearly identical. Hence, the total G appears to govern the delamination threshold for fatigue. Furthermore, comparison of the 5208 matrix composite to the H205 matrix composite shows a significant improvement in the static G_c, yet the magnitude of this improvement for the G_c threshold in fatigue is much less. A series of mixed-mode static and cyclic tests like these may be needed to evaluate toughened resin composites subjected to cyclic loads.

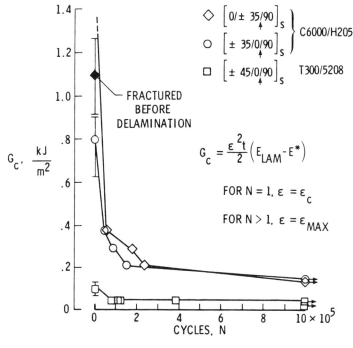

Figure 11

EDGE DELAMINATION TEST SPECIMEN SIZE OPTIMIZATION

As new matrix resins are developed, they must be evaluated for improved toughness. Ideally, this should be done on very small specimens because often only limited quantities of experimental resins are available. For this reason, a study was undertaken to miniaturize the edge delamination test. The configuration shown in figure 12 appears to be an optimal compromise between material and testing constraints. The G_c results from this specimen are identical to those measured on the larger coupons.

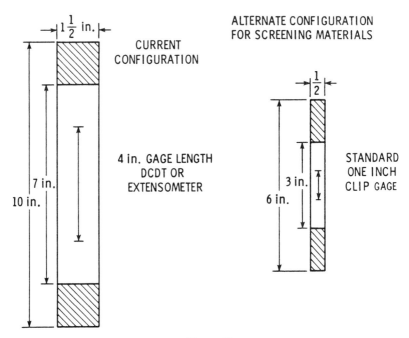

Figure 12

SUMMARY

A study was conducted to characterize the delamination resistance of toughened resin composites. Both the edge delamination test (EDT) and the double-cantilever-beam (DCB) test provided a useful ranking of improvements in delamination resistance between brittle and tough resin composites. Several layups were designed for the edge delamination test to cover a wide range of mixed-mode conditions. The DCB and the various layups of the EDT were then used to characterize the interlaminar fracture behavior of brittle and toughened resin composites subjected to both static and cyclic loading.

REFERENCES

1. O'Brien, T. K.; Johnston, N. J.; Morris, D. H.; and Simonds, R. A.: A Simple Test for the Interlaminar Fracture Toughness of Composites. SAMPE Journal, vol. 18, no. 4, 1982, pp. 8-15.

2. O'Brien, T. K.: Characterization of Delamination Onset and Growth in a Composite Laminate. Damage in Composite Materials, K. L. Reifsnider, ed., ASTM STP 775, American Society for Testing and Materials, 1982, pp. 140-167. (Also NASA TM-81940, January 1981)

3. Standard Tests for Toughened Resin Composites. NASA RP-1092, 1982. (Revised July 1983)

4. Raju, I. S.; and Crews, J. H., Jr.: Interlaminar Stress Singularities at a Straight Free Edge in Composite Laminates. Journal of Computers and Structures, vol. 14, no. 1-2, 1981, pp. 21-28.

5. O'Brien, T. K.: Mixed-Mode Strain Energy Release Rate Effects on Edge Delamination of Composites. NASA TM-85492, January 1983. (To appear in Effects of Defects in Composite Materials, D. J. Wilkins, ed., ASTM STP 836, 1984)

6. O'Brien, T. K.; Johnston, N. J.; Morris, D. H.; and Simonds, R. A.: Determination of Interlaminar Fracture Toughness and Fracture Mode Dependence of Composites Using the Edge Delamination Test. Proceedings of the International Conference on Testing, Evaluation, and Quality Control of Composites, TEQC83, T. Feest, ed., Butterworth Scientific Ltd., Kent, England, 1983, pp. 223-232.

COMPOSITE MATERIALS CHARACTERIZATION AND DEVELOPMENT AT AFWAL

C.E. Browning

Air Force Wright Aeronautical Laboratories
Wright-Patterson Air Force Base, Ohio

The first subject area to be discussed is Composite Characterization for Matrix Dominated Failure Modes. The objective of this work is to develop test methodology for characterizing matrix dominated failure modes, emphasizing the two major issues of matrix cracking and delamination under static loading. Also of major importance is establishing the relationship of the composite properties to the matrix properties.

OBJECTIVE - TO DERIVE TEST METHODOLOGY FOR CHARACTERIZING MATRIX DOMINATED COMPOSITE FAILURE MODES WITH EMPHASIS ON -

- MATRIX CRACKING
- DELAMINATION UNDER STATIC LOADING
- RELATING COMPOSITE PROPERTIES TO MATRIX RESIN PROPERTIES

The major characterization methodologies which are applicable to the characterization of matrix dominated composite failure modes are underlined below. Strength characterization entails a detailed stress analysis with a failure criterion to predict and measure experimentally the onset of matrix cracking and delamination. Fracture mechanics characterization entails the application of classical techniques of linear elastic fracture mechanics. We have actively worked all of the areas shown below except for impact characterization.

STRENGTH CHARACTERIZATION

- NEAT RESIN
    - TENSILE STRESS - STRAIN RESPONSE
    - SHEAR STRESS - STRAIN BEHAVIOR

- COMPOSITE
    - TRANSVERSE TENSION
    - LAMINATE IN-SITU TRANSVERSE PLY FAILURE
    - INPLANE SHEAR
    - INTERLAMINAR SHEAR
    - COMPRESSION STRENGTH

FRACTURE MECHANICS CHARACTERIZATION

- NEAT RESIN
    - CENTER-NOTCH TEST
    - EDGE-NOTCH TEST
    - COMPACT TENSION TEST

- COMPOSITE
    - DOUBLE CANTILEVER BEAM TEST
    - TRANSVERSE CENTER-NOTCH TEST
    - FREE-EDGE DELAMINATION TEST

IMPACT CHARACTERIZATION
FRACTOGRAPHY

With respect to strength characterization, the tests shown below have been investigated. The tests from this group to be discussed include neat resin tensile, composite transverse tension, laminate in-situ transverse ply failure, and composite interlaminar shear.

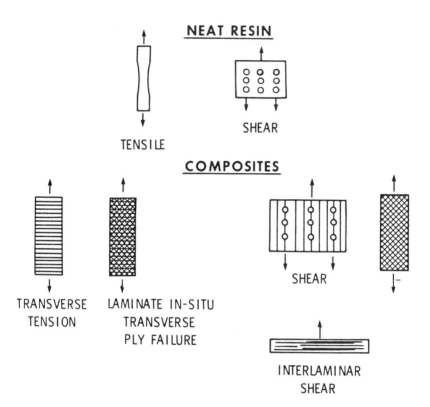

Neat resin tensile results are illustrated below. AF-R-E350 is a 350°F epoxy used in-house at the Materials Lab (ML). It contains MY-720 and DDS. This system typifies a brittle resin. Polysulfone is a thermoplastic that typifies a ductile resin. Typical stress strain responses are shown. The polysulfone curve actually continues to higher strain levels than shown.

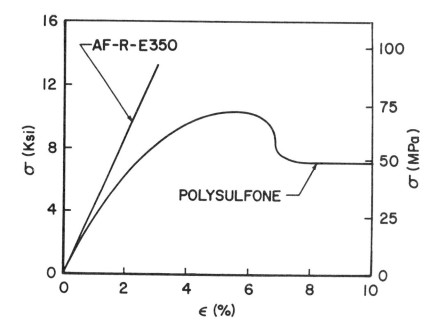

Composite transverse tension results are shown below for AF-R-E350 with AS-1 fiber at two fiber volume levels and polysulfone with AS-1 fiber. In many cases transverse tension is the limiting basis for design purposes. If the transverse ply failure is the limiting failure criterion, the transverse ply in a laminate will determine its strength. The data in the chart was normalized for fiber volume. It is important to normalize the transverse data as well as the unidirectional data. This was done by calculating strain concentration factors and using the Halpin-Tsai equations to extrapolate the epoxy to a fiber volume of 54%. This doesn't affect the stress but does change the strain as shown. Comparing the stress-strain curves for equivalent fiber volume shows that the strain-to-failure for the polysulfone/AS-1 composite vs. the epoxy/AS-1 is not in proportion to the neat resin data.

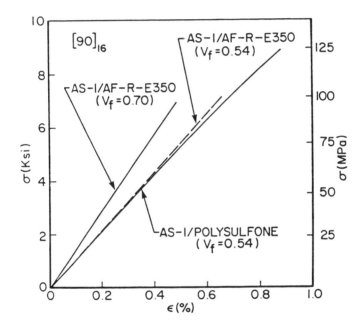

This chart shows that there may be a limiting value of composite strain-to-failure irrespective of neat resin properties. In the very ductile matrices, yielding of the neat resin during tensile testing leads to "necking" in the specimen at high strain levels. In the composite, due to the presence of the fibers, this necking is prevented, resulting in the resin being under a biaxial state of stress. Biaxial stress-strain behavior of a neat resin may be a better indicator of resin performance in the composite than uniaxial behavior.

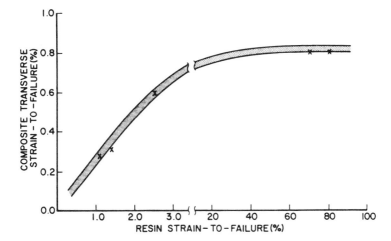

While a 90° tensile test gives useful information with respect to the transverse strain capability of a material, a multi-directional laminate containing 90° plies may be a more realistic approach to measuring and defining in-situ first-ply failure. A 90° tensile test provides initial transverse ply failure information only, while a multi-directional laminate containing multiple 90° plies allows one to observe multiple cracks in multiple transverse plies to assess their total effect on laminate mechanical behavior. Other plies in a multi-directional laminate can increase 90° ply failure strain due to constraints, resistance to crack growth, etc. This phenomenon is illustrated below for a (±45/90) laminate. Specimen edges are polished. Cracks in 90° plies are counted as a function of tensile testing. The knee in the stress-strain curve corresponds to the increase in crack density.

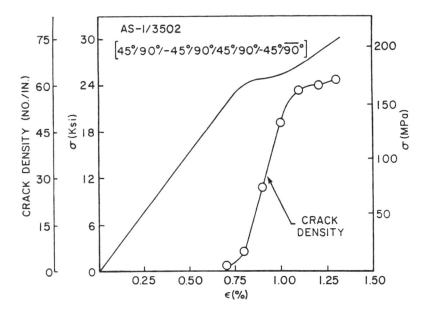

Another type of strength characterization test is an interlaminar (I-L) beam test. Three types of I-L beam tests investigated at the ML are shown below. The top one is the conventional short-beam shear specimen tested in 3-point loading with a span-to-depth ratio (L/h) of 4. Because this specimen rarely produces a shear failure, the other two coupons were investigated. One alternative is a thick (50 plies) beam tested in 3-point bending. This specimen consistently gives shear failures; however, its thickness creates undesirable processing and material problems. A better alternative is the four-point shear test shown at the bottom. This specimen consistently gives shear failures.

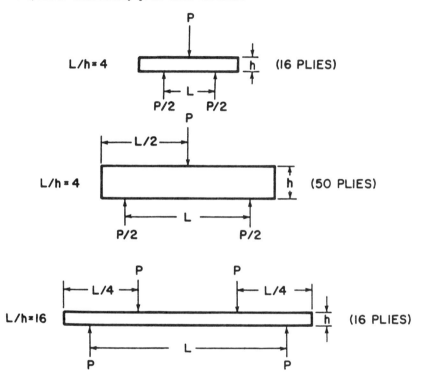

36 Tough Composite Materials

Shown below are photomicrographs of failed specimens representative of the three specimens on the previous figure. They provide additional insight into the complexity of beam failure modes. The failed 16-ply, 3-point beam specimen typically shows localized buckling adjacent to the top loading pin (compression surface). In the case of the 50-ply, 3-point beam, a vertical crack occurs under the top loading pin. This crack eventually leads to a horizontal shear failure. The 16-ply, 4-point beam specimen (bottom figure) also exhibits a vertical crack initially under the top loading pin. This crack eventually creates a horizontal shear crack.

16-PLY, 3-POINT BEAM

50-PLY, 3-POINT BEAM

16-PLY, 4-POINT BEAM

The beam failures occur in regions where the stress distribution is very complex. The shear stress distribution at three different regions of a 16-ply beam are shown below. Classical beam theory is obtained only over a limited segment of the beam. Otherwise there will be a very complex stress distribution with high stress concentrations under the loading pins. Complex failure modes and complex stress distributions make I-beam specimens very difficult to interpret. The usefulness of I-beam tests are of questionable value.

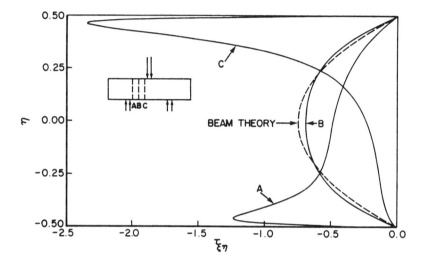

Tough Composite Materials

The next major area of characterization is Fracture Mechanics of Composites. The test methods shown below have been investigated. With respect to composites, major emphasis has been on interlaminar fracture mechanics. This seems to be a natural approach to treating delamination.

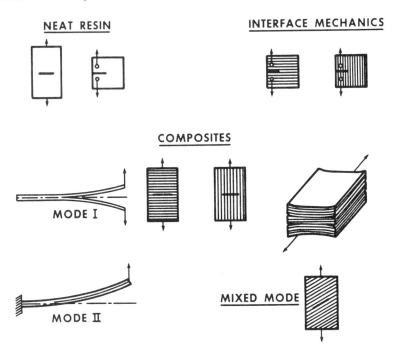

Composite Fracture Toughness and Impact Characterization

With respect to neat resin fracture toughness characterization, the three configurations shown below have been investigated with both dull and sharp cracks. The results show that the compact tension specimen on the right provided the best data with respect to resin vs. composite properties for the cases of the two epoxies. A major discrepancy exists with the polysulfone which may be due to the effective size of the plastic zone at the crack tip.

$$G_{Ic}, \text{ in-lb/in}^2 \ (10^{-2} \text{ J/cm}^2)$$

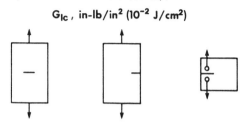

MATERIAL	DULL CRACK	SHARP CRACK	DULL CRACK	SHARP CRACK	DULL CRACK	SHARP CRACK	COMPOSITE DCB
BRITTLE EPOXY	6.5 (11.4)	-	4.1 (7.2)	0.5 (0.9)	-	0.8 (1.4)	0.8 (1.4)
DUCTILE EPOXY	12.1 (21.3)	-	-	-	24.0 (42.2)	0.6 (1.1)	0.8 (1.4)
POLYSULFONE	11.7 (20.6)	-	11.0 (19.4)	28.0 (49.3)	32.0 (56.3)	15 (26.4)	3.7 (6.5)

Most of our characterization work has been with composites. The principal technique used is the double cantilever beam (DCB) test shown below. The DCB coupon is straight-sided, 9-in. long, 1-in. wide, 24 plies, 0°, with a starter crack. Tests are run over ½-in. increments. Data reduction is by the area method. Test data will be shown in a subsequent chart.

MODE I TEST FOR INTERLAMINAR TOUGHNESS

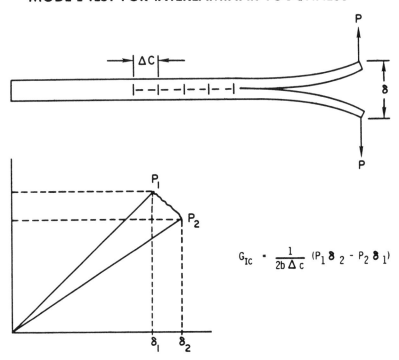

$$G_{IC} = \frac{1}{2b \Delta c} (P_1 \delta_2 - P_2 \delta_1)$$

Composite Fracture Toughness and Impact Characterization

A potential alternative to the DCB test is the transverse center-notch (CN) test. This specimen uses three crack sizes with three specimen widths. Results show it to be a reasonable alternative. Data will be shown in subsequent chart.

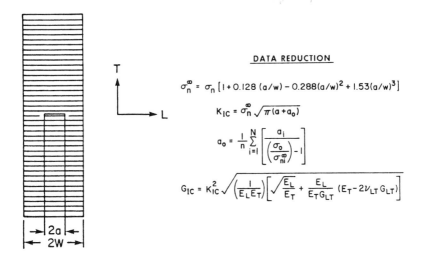

DATA REDUCTION

$$\sigma_n^\infty = \sigma_n \left[1 + 0.128 (a/w) - 0.288(a/w)^2 + 1.53(a/w)^3 \right]$$

$$K_{IC} = \sigma_n^\infty \sqrt{\pi(a+a_o)}$$

$$a_o = \frac{1}{n} \sum_{i=1}^{N} \left[\frac{a_i}{\left(\frac{\sigma_o}{\sigma_{ni}^\infty}\right) - 1} \right]$$

$$G_{IC} = K_{IC}^2 \sqrt{\left(\frac{1}{E_L E_T}\right) \left[\sqrt{\frac{E_L}{E_T}} + \frac{E_L}{E_T G_{LT}} (E_T - 2\nu_{LT} G_{LT}) \right]}$$

Another important test is the free edge delamination (ED) test. The stacking sequence is chosen such that delamination is induced along the straight-sided free edge under tension loading. The test discriminates between brittle and ductile resins and is an aid in screening matrix resins for toughness.

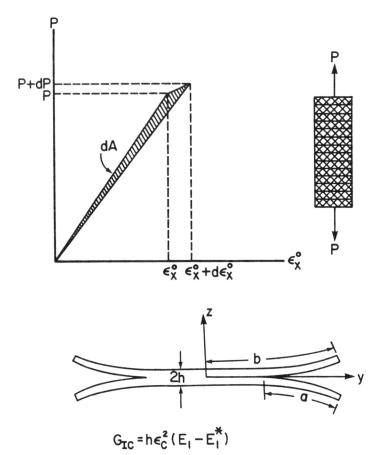

$$G_{IC} = h\epsilon_c^2 (E_I - E_I^*)$$

Composite Fracture Toughness and Impact Characterization

This table summarizes the results of Mode I delamination tests for the methods previously discussed as a function of several materials variables. The data shows that the DCB test is the preferred tool for characterizing interlaminar Mode I energy release rates.

MATERIAL	V_f	G_{Ic} in lb/in^2 (10^{-2} J/cm^2)		
		DCB	90° CN	ED
AS-1/AF-R-E350	0.70	.752 (1.31)	1.32 (2.31)	1.11 (1.94)
AS-1/3502	0.70	.801 (1.40)	.881 (1.54)	1.53 (2.67)
AS-1/POLYSULFONE	0.54	3.74 (6.55)	---	---
AS-1/ATQ	0.55	2.10 (3.68)	---	.170 (.298)
AS-1/3502 $[90_2^0/0_8^0/90_2^0]_s$	0.70	1.60 (2.80)	---	---
XAS/PEEK (APC-1)	0.55	8.00 (14.08)	---	8.00 (14.08)

The edge delamination coupon is very useful for screening resins. Data is shown below for a series of materials of different degrees of brittleness/ductility. The ratio of the delamination stress to the ultimate stress shows that the test can discriminate between matrix resins.

COMPOSITE EDGE DELAMINATION
$(+30_2/-30_2/90_2)_s$

MATERIAL	σ_{DEL} (KSI)	σ_{ULT} (KSI)
AF-R-E350/AS-1	30.0	35.5
V378A/T-300	21.0	24.0
ATQ/AS-4	9.0	25.0
PS/AS-1	NO DELAM	46.0
PEEK/XAS	NO DELAM	60.0

Composite Fracture Toughness and Impact Characterization   45

Important summary points are shown below.

- RESIN TENSILE/COMPOSITE TRANSVERSE TENSION ARE USEFUL STRENGTH TESTS
- INPLANE SHEAR LESS INFORMATIVE
- LAMINATE TEST FOR INTERROGATING IN-SITU TRANSVERSE STRENGTH SHOULD BE UTILIZED
- SHORT BEAM SHEAR TEST SHOULD BE DELETED
- COMPACT TENSION IS A VIABLE RESIN FRACTURE TEST. ATTENTION MUST BE GIVEN TO SPECIMEN GEOMETRY/ MATERIALS CHARACTERISTICS
- DOUBLE CANTILEVER BEAM TEST IS VIABLE MODE I - INTERLAMINAR FRACTURE TEST. SHOULD BE RESTRICTED TO $0°$ LAMINATES AT THE PRESENT TIME. ATTENTION MUST BE GIVEN TO SPECIMEN GEOMETRY/ MATERIALS CHARACTERISTICS.
- FREE-EDGE DELAMINATION TENSILE TEST IS A POTENTIAL USEFUL ALTERNATIVE TO THE DOUBLE CANTILEVER BEAM TEST. IS ALSO USEFUL AS AN INTERLAMINAR TENSILE STRENGTH TEST FOR MATRIX SCREENING.

Areas of particular interest for future work are summarized below.

- RELATIONSHIP BETWEEN MATRIX STRESS-STRAIN RESPONSE AND UNIDIRECTIONAL TRANSVERSE TENSION AND INPLANE SHEAR
- MATRIX STRESS-STRAIN RESPONSE UNDER BIAXIAL LOADING
- BETTER UNDERSTANDING OF SHEAR AS A FAILURE MODE
- MECHANISM OF IN-SITU TRANSVERSE PLY FAILURE - RELATIONSHIP TO MATRIX PROPERTIES
- VIABLE MODE II AND MIXED MODE INTERLAMINAR FRACTURE TESTS

Composite Fracture Toughness and Impact Characterization   47

The next major discussion topic is Composite Materials Development at the ML. In so far as the investment of resources is concerned, the major areas of interest are those shown.

MAJOR AREAS OF INTEREST:

- 350-450°F USE/EPOXY-REPLACEMENT TYPE
  - ACETYLENE-TERMINATED (AT) RESINS
  - BISMALEIMIDE (BMI) RESINS

- THERMOPLASTICS
  - PEEK
  - REACTIVE PLASTICIZER CONCEPTS

- FAILURE RESISTANT COMPOSITE CONCEPTS
  - LAMINATE CONSTRUCTION

An important resin technology is that of acetylene terminated (AT) resins. A variety of acetylene end-capped resins are attainable by tailoring the backbone structure. A range of properties, including use temperature, are attainable. Backbone structures possible include imides, sulfone, bis-phenols, etc.

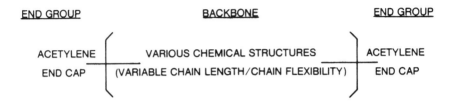

USE TEMPERATURE

LONG TERM   SHORT TERM

250 - 550°F   600 - 650°F

A specific example of an AT resin with high potential is the ATBA resin. It possesses the outstanding attributes listed. In addition this resin possesses excellent retention of hot/wet properties with low moisture uptake.

ACETYLENE-TERMINATED BISPHENOL A

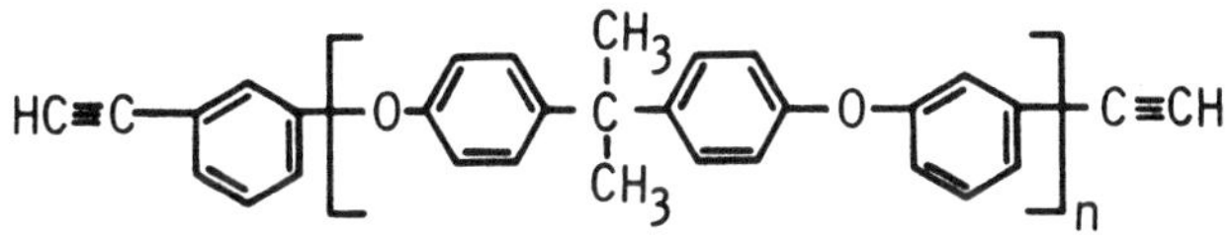

$$HC\equiv C-\phenyl-\left[O-\phenyl-C(CH_3)_2-\phenyl-O-\phenyl\right]_n-C\equiv CH$$

HANDLING/PROCESSING CHARACTERISTICS:

- SINGLE COMPONENT
- LONG SHELF-LIFE/OUT-TIME
- EPOXY-LIKE PROCESSING
  - HOT-MELT
  - TACK/DRAPE
  - AUTOCLAVE CURE-400°F/100 PSI/NO BAG
  - POSTCURE - 482°F

The performance of AT resins, an AT-bisphenol A and an AT-quinoxaline, are shown versus performance of typical epoxy systems.

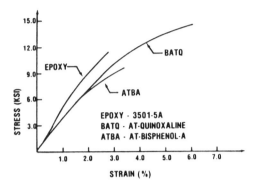

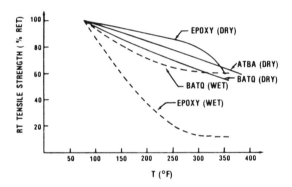

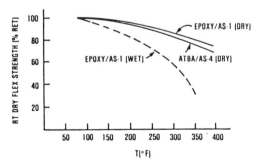

Another matrix resin technology of major importance is that of bismaleimides (BMI's). Areas to be addressed are shown in the chart.

AREAS OF ACTIVITY:

- IMPROVED PREPREG DEVELOPMENT (CONTRACT - FY 83)
  - PREPREG HANDLEABILITY
  - TOUGHNESS

- STRUCTURE - PROPERTY - PROCESSING RELATIONSHIPS (IN-HOUSE - FY 83)

- PROCESSING SCIENCE/QUALITY ASSURANCE (CONTRACT - FY 85)

Another major materials technology area is that of thermoplastics. Thermoplastics offer significant advantages over contemporary materials in two key areas - damage tolerance and processing. Major areas of activity are summarized below.

AREAS OF ACTIVITY:

- TECHNOLOGY DEVELOPMENT
  - TP COMPOSITE TECHNOLOGY DEVELOPMENT (CONTRACT - FY 83)
  - IMPROVED TP COMPOSITE MATERIAL FORMS / PROCESS DEVELOPMENT (CONTRACT - FY 84)

- MANUFACTURING METHODS FOR TP COMPOSITES (CONTRACT - FY 85)

- PROCESSING SCIENCE / QUALITY ASSURANCE (CONTRACT - FY 85)

- MORPHOLOGY (IN-HOUSE - FY 83)

- REACTIVE PLASTICIZER CONCEPTS DEVELOPMENT (IN-HOUSE - FY 83)

Composite Fracture Toughness and Impact Characterization

A considerable amount of work has been done on a semi-crystalline thermoplastic for advanced composites - polyetheretherketone (PEEK) from ICI. The results shown below provide the impetus to fully exploit the benefits thermoplastics have to offer.

POLYETHERETHERKETONE (PEEK)

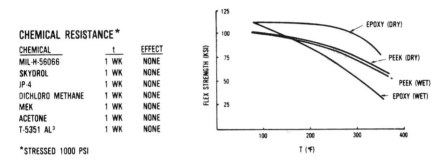

CHEMICAL RESISTANCE*

CHEMICAL	t	EFFECT
MIL-H-56066	1 WK	NONE
SKYDROL	1 WK	NONE
JP-4	1 WK	NONE
DICHLORO METHANE	1 WK	NONE
MEK	1 WK	NONE
ACETONE	1 WK	NONE
T-5351 AL³	1 WK	NONE

*STRESSED 1000 PSI

- IMPROVED TOUGHNESS (G_{1c} 11.4 VS 1.4 LBS/IN)
- INDEFINITE RT SHELF LIFE
- REDUCED QA COSTS
- REDUCED SCRAPPAGE
- LOWER COST FABRICATION

EFFECT OF IMPACT DAMAGE AND OPEN HOLES ON THE COMPRESSION STRENGTH OF TOUGH RESIN/HIGH STRAIN FIBER LAMINATES

Jerry G. Williams
NASA Langley Research Center
Hampton, Virginia

INTRODUCTION

Past experience has shown that structural damage and design-based inclusions such as cutouts can significantly reduce the strength of graphite-epoxy laminates (ref. 1). One composite mechanics research activity at the Langley Research Center is to assess and improve the performance of composite structures damaged by impact or containing local discontinuities such as cutouts. Reductions in strength are common to both tension and compression loaded laminates; however, the problem associated with compression performance has been found to be the most elusive to solve. Compression failure involves both shear crippling and delamination modes. Small-scale coupon tests have not yet been developed to adequately predict damaged-laminate compression performance reductions. Two plate specimen configurations, however, have been developed by NASA (ref. 2) to help define the severity of the compression strength reduction problem and to assess the relative merit of proposed toughened material systems. These two test configurations, one involving impact damage and the other open hole specimens, are shown in figure 1. The test technique for impact specimens involves damaging the plate at selected energies, measuring the size of damage by ultrasonic C-scan techniques and measuring the residual strength in a compression load test. Open-hole specimen compression tests are conducted for several different hole diameters and the failure strain and load and mode of failure recorded. The plate specimen used in these tests is designed with length, width, thickness and laminate stiffness to ensure that overall plate buckling is not responsible for initiating failure.

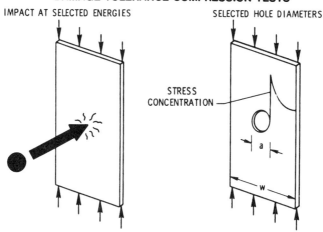

Figure 1

TEST SPECIMENS AND FIXTURE

In the current investigation, several new graphite-epoxy material systems proposed for improved damage tolerance and listed in figure 2 were studied. Material parameters included both tough resin formulations and high strain fibers. Material suppliers included Narmco (T300 fiber and 5208 resin), American Cyanamid (BP907 resin), and Hercules (AS4 and AS6 fibers and 3502, 2220-1, and 2220-3 resin). Ultimate tensile strains for these fibers are approximately: 1.2% - T300, 1.4% - AS4 and 1.8% - AS6. The T300/5208 material is used as a baseline and T300/BP907 was identified in past sudies as exhibiting improved damage tolerance characteristics (ref. 3). All tests were conducted at room temperature and therefore do not address the reduction in strength of resin materials such as BP907 caused by moisture and elevated temperatures. Quasi-isotropic laminate specimens approximately 0.25 inches thick and 10 inches long by 5 inches wide were tested in the fixture shown in figure 2. The fixture imposed nearly clamped boundary conditions on the loaded ends and simple support boundary conditions on the lateral edges. Two sets of strain gages mounted back-to-back were used to measure the axial strain.

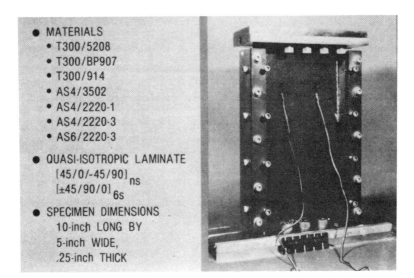

- MATERIALS
  - T300/5208
  - T300/BP907
  - T300/914
  - AS4/3502
  - AS4/2220-1
  - AS4/2220-3
  - AS6/2220-3

- QUASI-ISOTROPIC LAMINATE
  $[45/0/-45/90]_{ns}$
  $[\pm 45/90/0]_{6s}$

- SPECIMEN DIMENSIONS
  10-inch LONG BY
  5-inch WIDE,
  .25-inch THICK

Figure 2

LAMINATE DAMAGE FOLLOWING IMPACT

An illustration of the influence the resin material has on the size and extent of damage in a graphite-epoxy laminate resulting from projectile impact is shown in figure 3. The damage following impact by a 1/2-inch diameter aluminum sphere at approximately 13 ft-lb of energy is shown on the top row for a brittle behavior resin and on the bottom row for a toughened resin system. The orthotropic laminate is approximately 0.25 inch thick. Less damage is observed for the tough resin material by visual observation of surface damage, by ultrasonic C-scan inspection and by microscopic inspection of a cross-section through the impact damage zone. This demonstration shows, therefore, that it is possible to tailor the matrix material properties to reduce the size of damage following projectile impact.

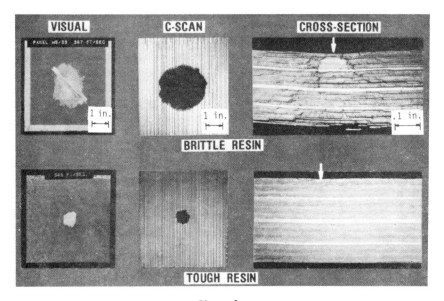

Figure 3

EFFECT OF IMPACT ENERGY ON DAMAGE SIZE

A plot of the damage area measured using ultrasonic C-scan signatures for several material systems is presented in figure 4 as a function of the projectile impact velocity and energy. The threshold energy at which damage can first be detected varies for the materials studied; however, all are in the range of three to five ft-lb. The largest damage size was measured for the T300/914 material. One variable for this material that differed from the other materials was that a thicker prepreg tape was used, resulting in approximately half as many plies for the 0.25-inch thick laminate. The effect of lamina thickness is not established. The results of several of the materials fall within a relatively narrow band for the energies studied. There appears to be a divergence of the results, however, at the upper energy levels, a trend which merits further study.

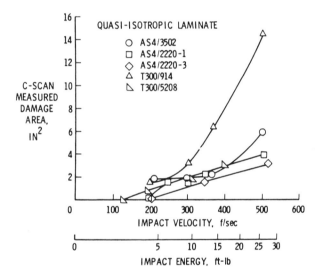

Figure 4

IMPACT INITIATED COMPRESSION FAILURE MODES

Experimental studies have shown that the failure of damaged composite laminates loaded in compression involves two primary failure mechanisms: delamination and transverse shear (ref. 4). These two failure mechanisms are illustrated in figure 5 for a brittle resin laminate and for a damage-tolerant tough resin laminate. The photographs on the right of the figure show cross-sections of failure regions which are typical for these two classes of material. The brittle resin laminate shows considerable evidence of delamination whereas the tough resin laminate cross-section is characterized by a through-the-thickness shear band which is approximately 0.07 inch wide. Closer inspection reveals, however, that both specimens actually exhibit both delamination and transverse shear failure mechanisms. The transverse shear failure mode for the brittle resin laminate develops in only a few plies before delamination occurs, while the transverse shear mode for tough resin laminates is several plies thick before it is interrupted by delamination caused by wedges of failed material prying apart the plies. Tough resin formulations improve damage tolerance by suppressing the delamination mode of failure, permitting failure to occur at the next higher energy mode involving transverse shear.

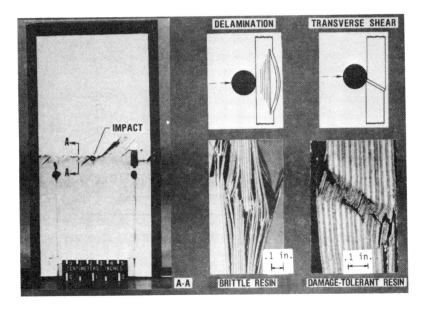

Figure 5

Composite Fracture Toughness and Impact Characterization

PROPAGATION OF IMPACT INDUCED DELAMINATION

The sequence of events which occurs when a brittle resin is damaged by impact and subsequently loaded in compression to failure is shown in figure 6. The moire fringe photographs show the local out-of-plane deformations of the laminate in the impact damaged region. Photographs presented left to right correspond to increasing load up to ultimate at which damage propagates from the center of the panel to the two lateral edges. Sublaminates caused by impact-induced delaminations have reduced bending stiffnesses compared to the undamaged laminate and, if sufficiently large, buckle at significantly lower loads than the overall plate buckles. These local buckles represented by the moire fringe contours cause high stresses in the resin at the delamination boundary. When the buckle is sufficiently advanced, these stresses cause fracture of the resin and the damage propagates.

Figure 6

SHEAR CRIPPLING FAILURE MODE

The shear crippling mode of failure occurs not only at the macroscopic scale as illustrated in figure 5 but also on the microscopic scale involving individual graphite fibers as illustrated in figure 7. This tough resin orthotropic laminate was damaged by impact and loaded until the damage began to propagate across the panel. The damage propagation arrested; the load was removed and a cross-section was taken through the damaged region. Shown on the right of figure 7 is a photomicrograph of four of the interior plies $[45/0_2/-45]$ of the 48-ply laminate. Graphite fibers in the zero degree plies (aligned coincident with the applied load) failed by shear crippling while fibers oriented at 45 degrees were undamaged. The model proposed to explain this phenomenon is that the strain concentration in zero-degree plies located in the damage zone and the reduced support to the fibers due to matrix fracture cause the graphite fibers to microbuckle. Fracture of the fiber occurs when the axial plus postbuckling bending strains reach a critical value.

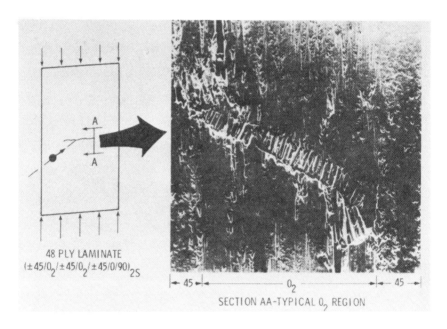

Figure 7

FAILURE OF OPEN-HOLE SPECIMEN LOADED IN COMPRESSION

A series of photographs showing the initiation and propagation of delamination for an open hole specimen loaded in compression is shown in figure 8. At 95.2% of the ultimate load, moire fringe photographs show no evidence of delamination around the hole boundary. At 95.4%, local fringes appear and grow in size with increasing load as can be seen comparing the photographs at 95.9% and 98.1% of ultimate. Ultimate failure occurs when damage propagates completely across the reduced section of the plate. One might conclude based on this evidence that the initiating failure mode for open hole specimens is delamination; however, as will be shown in the next figure, microscopic shear crippling occurs in the vicinity of the hole boundary in advance of delamination.

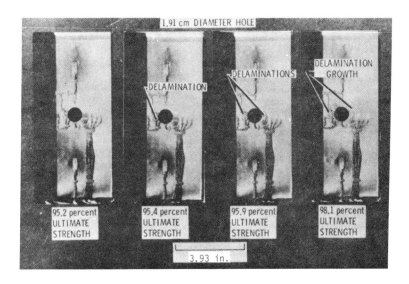

Figure 8

SHEAR CRIPPLING INITIATES OPEN-HOLE SPECIMEN FAILURE

Another specimen similar to the one shown in figure 8 was loaded to a load level just prior to the initiation of delamination (approximately 92% of ultimate) and unloaded. A small block of material adjacent to the hole boundary was cut from the specimen and surface material sanded away to expose an interior 0-degree layer. Scanning electron photomicrographs of this region are shown on the right of figure 9. Damage is the same failure of individual graphite fibers by shear crippling which was shown earlier in figure 7 for the compression failure of impact-damaged laminates. The higher magnification photomicrograph shows the failed fiber length to diameter ratio to be approximately four. The proposed failure model is the same as proposed earlier, i.e., graphite fibers microbuckle in the high strain concentration region adjacent to the hole and fail in the post-buckled state.

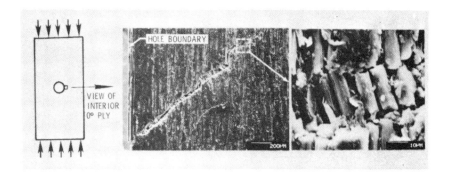

Figure 9

IMPACT-DAMAGE FAILURE THRESHOLD CURVE

The effect of impact damage on the failure strain of a compression loaded graphite-epoxy laminate constructed with a brittle-behavior resin material is presented in figure 10 (ref. 4). Filled circular symbols represent specimens which failed catastrophically when loaded to the indicated strain level and impacted by a 1/2-inch diameter aluminum sphere at the indicated velocities. Open circular symbols represent specimens which may have been damaged by impact, but the damage was contained with little loss of load. A narrow band separates open and closed symbols and a failure threshold curve has been drawn through the band, thus separating the graph into two zones. Impact conditions above and to the right of the curve result in specimen failure while the laminate survived the less severe conditions below and to the left of the curve. A severe reduction in strength occurs for impacts in the 165 to 250 ft/sec range and at 330 ft/sec the failure threshold strain is reduced to approximately 0.0028.

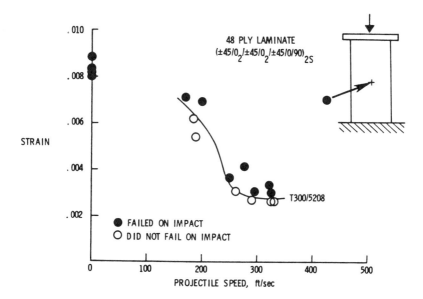

Figure 10

TOUGH RESIN IMPROVES LAMINATE DAMAGE TOLERANCE

Failure threshold curves for a 48-ply orthotropic graphite-epoxy laminate constructed using two different resin systems are shown in figure 11 (ref. 4). For the test conditions studied, the tough BP907 resin system shows substantial improvement relative to the brittle 5208 resin system. Similar improvements have also been observed for a 5208 resin laminate when it was reinforced by through-the-thickness stitching. The explanation for this improvement is that both the tough resin system and stitching suppress the delamination mode of failure. The delamination mode of failure has been studied using fracture toughness tests such as the double cantilever beam. Improved compression strength after impact has been correlated with fracture toughness measurements for material systems with widely varying fracture toughness properties such as the materials compared in figure 11. As shown in figure 7, however, shear crippling is also involved in the failure of impact-damaged laminates and fracture toughness tests do not address this mode of failure.

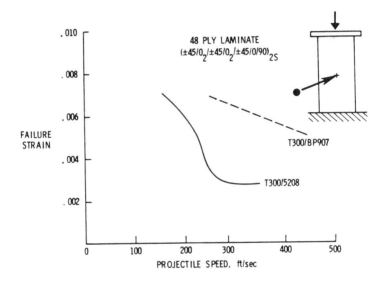

Figure 11

EFFECT OF SIZE OF IMPACT DAMAGE ON FAILURE STRAIN

A need exists for a comparison method which the composite structure designer can use to assess the effect of various impact conditions and material systems on structural strength. Trends for test data in which increasing strength losses were observed to occur with increasing impact damage size suggested the parameters used in the graph presented in figure 12. The failure strain for several different material systems constructed in a quasi-isotropic laminate is plotted as a function of the width of damage resulting from impact. The damage width was determined from ultrasonic C-scan photographs and is normalized by the specimen width (5 inches). For these laminates and impact conditions, the size of damage appears to be a parameter that reduces the test data for all four materials to a common curve. A two-parameter curve asymptotic to a/w = .24 has been drawn through the data. A large reduction in strength occurs around a/w = .24 and the failure strain for a/w < .24 is governed by conditions other than impact such as plate buckling. Additional study is required to assess the generalization of this data to other impact conditions and laminates.

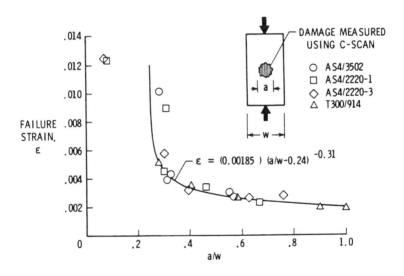

Figure 12

OPEN-HOLE COMPRESSION SPECIMENS

A series of 5-inch-wide and 10-inch-long quasi-isotropic specimens with selected centrally located holes were tested for several different material systems. Photographs of some of these specimens are presented in figure 13.

Figure 13

STRESS-STRAIN RESPONSE FOR OPEN-HOLE SPECIMENS

The stress-strain response up to failure for AS4/2220-3 quasi-isotropic 5-inch-wide specimens with selected a/w hole sizes is presented in figure 14. Strain data is taken from strain gages located near one end of the specimen. For large holes, the stress-strain response deviates from the no-hole (a/w = 0) curve. For purposes of data comparison, the failure strain for open-hole specimens reported in subsequent figures is the strain which the no-hole specimen carried at the same stress that the open-hole specimens carried at failure.

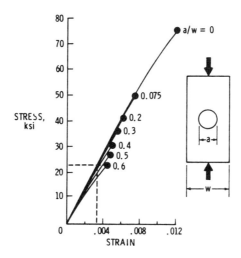

Figure 14

EFFECT OF CIRCULAR HOLES ON COMPRESSION STRENGTH

A comparison of the reduction in strength for a brittle (T300/5208) and tough (T300/BP907) resin system laminate is presented in figure 15 as a function of the hole diameter "a" normalized by the specimen width "w" (ref. 4). The curve faired through the data is a failure prediction base on the point-stress failure criterion proposed by Whitney and Nuismer (ref. 5). The curve is bounded on the top by a net-area notch-insensitive curve and on the bottom by a notch-sensitive curve in which failure is assumed to occur when the stress at the hole edge reaches the critical value for an unnotched specimen. The different resin formulations appear to have had no effect on the failure strain for these two orthotropic laminates. The explanation for this apparent paradox in which the tough resin improved the strength of impact damaged specimens (fig. 11) but not open-hole specimens involves understanding the governing failure mechanisms. For impact damage, tough resins improved the performance by suppressing the delamination mode of failure. For open hole specimens, as shown in figure 9, the failure initiation mechanism involves fiber microbuckling and shear crippling of highly stressed material adjacent to the hole. Fiber microbuckling is governed by the stiffness properties of the matrix and fiber and by other factors such as the integrity of the matrix-to-fiber bond. Similar strength reductions for these two material systems with holes occur since the same fiber was used in both laminates and because the two resin systems have similar initial elastic modulus properties.

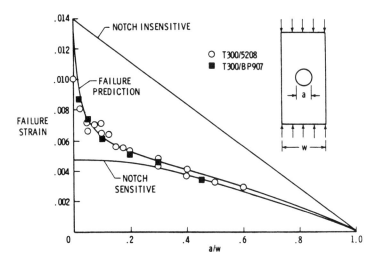

Figure 15

EFFECT OF HIGH STRAIN FIBER ON FAILURE STRAIN OF OPEN-HOLE SPECIMENS

The influence of hole size "a/w" on the failure strain of several quasi-isotropic laminates constructed with selected resin systems and two graphite fiber materials is presented in figure 16. The two theoretical failure curves are point stress failure predictions with the indicated characteristic parameters. The lower theoretical curve is taken from reference 6 and represents the best fit to date for T300/5208 graphite-epoxy. The data appear to group according to fiber reinforcement type with the AS4 fiber laminates exhibiting higher failure strain than T300 fiber laminates. The ultimate tension strains for T300 and AS4 are approximately 0.012 and 0.015, respectively. Recall from figure 15 that laminates with two different resin systems and the same fiber had identical strengths. If, as hypothesized, high bending strains in a buckled fiber initiate local failure, then one might expect a higher tension strain fiber to exhibit a higher laminate strength as was observed in this series of tests. Several material and structural properties govern fiber microbuckling and failure including the fiber extensional and bending stiffness and strength and the stiffness and strength properties of the matrix. Theoretically, a high shear modulus property of the resin should also increase the strain at which microbuckling would occur. All of the factors which affect microbuckling and compression strength need to be better understood in order to better tailor material and laminate properties for optimum performance.

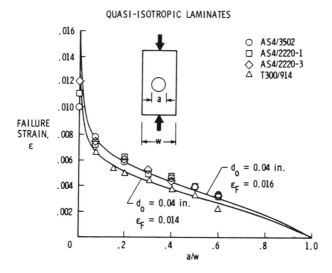

Figure 16

OPEN-HOLE VERSUS IMPACT STRENGTH REDUCTION

Designers of composite structures must address the effects of both holes and impact on design allowables and are interested in the range of conditions in which each factor governs structural performance. A comparison is made in figure 17 of the effect of these two types of local discontinuities. The open hole curves are taken from figure 16 for AS4 and T300 fiber laminates and the impact curve is taken from figure 12 in which the damage size was determined from C-scan measurements. The open hole causes the greatest reduction in strength for $a/w \leq .3$ (w = 5 inches) and impact damage causes the greatest reduction for $a/w > .3$. Additional curves need to be defined for other plate widths and laminates to establish the generality for design purposes of these findings.

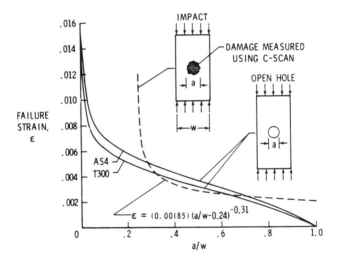

$$\varepsilon = (0.00185)(a/w-0.24)^{-0.31}$$

Figure 17

CONCLUSIONS

1. Tough resin system can reduce the size of the damage zone caused by impact.

2. Delamination and shear crippling are two fundamental mechanisms involved in the compression failure of graphite-epoxy laminates.

3. Tough resins (compared to brittle resins) can improve the compression strength of impact-damaged laminates by suppressing the delamination mode of failure.

4. Tough resins do not provide similar improvement in the performance of laminates with open holes where shear crippling is the dominant failure mechanism.

5. Several graphite-epoxy material systems were found to exhibit common strength reductions for equal size impact damage.

6. Higher strain fiber provided increase in failure strain of open hole specimens.

REFERENCES

1. Starnes, James H., Jr.; Rhodes, Marvin D.; and Williams, Jerry G.: Effect of Impact Damage and Holes on the Compressive Strength of a Graphite-Epoxy Laminate. ASTM STP 696, pp. 145-171, 1979.

2. Standard Tests for Toughening Resin Composites. NASA RP-1092, July 1983.

3. Williams, Jerry G.; and Rhodes, Marvin D.: Effect of Resin on the Impact-Damage Tolerance of Graphite-Epoxy Laminates. ASTM STP 787, pp. 450-480, 1983.

4. Starnes, James H., Jr.; and Williams, Jerry G.: Failure Characteristics of Graphite-Epoxy Structural Components Loaded in Compression. Mechanics of Composite Materials - Recent Advances. Hashin and Herakovich, Editors, Pergamon Press, 1983.

5. Whitney, J. M.; and Nuismer, R. J.: Stress Fracture Criteria for Laminated Composites Containing Stress Concentrations. Journal of Composite Materials, Vol. 8, July 1974, pp. 253-265.

6. Rhodes, Marvin D.; Mikulas, Martin M., Jr.; and McGowan, Paul E.: Effect of Orthotropic Properties and Panel Width on the Compression Strength of Graphite-Epoxy Laminates With Holes. AIAA Paper 82-0749 presented at AIAA/ASME/ASCE/AHS 23rd Structures, Structural Dynamics and Materials Conference. May 10-12, 1982.

EFFECTS OF CONSTITUENT PROPERTIES ON COMPRESSION FAILURE MECHANISMS

H. Thomas Hahn

Washington University
St. Louis, Missouri

INTRODUCTION

Recent work by Williams and Rhodes [1] has shown that higher impact resistance can be obtained for composites by using a tough resin because a tough resin can absorb more energy and localize the impact damage. However, a tough resin usually cannot be obtained without a sacrifice on modulus, and the lower modulus may lead to a lower compressive strength for composites. Therefore, a judicious selection of resin should be based on a balanced evaluation of both impact resistance and compressive strength. Although the subject of compression failure of composites has been frequently addressed in the literature, how constituent phases contribute to the overall compression strength still needs to be elucidated. Thus, the objectives of the present work are to delineate compression failure mechanisms in unidirectional composites and to identify material parameters that control compressive strength. With this enhanced understanding, laminate behavior can finally be predicted from lamina behavior.

BACKGROUND

- Higher impact resistance requires tougher resin
- Tougher resin may reduce compressive strength

OBJECTIVES

- Delineate compression failure mechanisms
- Identify material parameters that need improvement
- Correlate lamina behavior with laminate behavior

APPROACH

Compression failure mechanisms are studied at three different levels of material construction. The first level is a fiber bundle embedded in epoxy. Using this type of specimens, one can monitor the sequence of failure of the fiber bundle because the bundle is well contained within the epoxy. Also, the failure modes of fibers themselves can be studied at this level. The second level is unidirectional laminas loaded in the fiber direction. A comparison between the results from these two levels will help us understand the influence on compressive strength of the internal structure of composites. Multidirectional laminates, reserved for the last level, are used to assess the effects of off-axis plies and interfaces between plies. Different combinations of fibers and matrices are used because one failure mechanism operating in one material system may be suppressed in another material system. The present paper is concerned with the behavior of fiber bundle specimens only. The work on unidirectional and multidirectional laminates is still in progress. The IITRI compression fixture was used with a gage length of 13 mm. Specimens varied from 4 to 6.5 mm in thickness while being held at the same width of 13 mm. An Instron testing machine was used at a cross-head speed of 1.3 mm/min. During testing the fiber bundle was monitored for failure through a microscope at magnifications up to 50X. Since the fiber bundle occupied a small fraction of the specimen volume, the bundle failure did not lead to the specimen failure. Thus the bundle failure could be contained and monitored.

APPROACH

- Study compression failure of
  Fiber bundle in epoxy
  Unidirectional lamina
  Laminate
- Different fibers and matrices

EXPERIMENTAL PROCEDURE

- 4 - 6.5 x 6.4 x 13 mm
- IITRI compression fixture
- 1.3 mm/min.
- Fiber failure monitored through a microscope at up to 50X magnification

MATRIX AND FIBER PROPERTIES

Representative matrix and fiber properties are listed in the tables below. Two different epoxies were used: Epon 828 with curing agent Z and Epon 815 with curing agent V140. The former epoxy is more brittle than the latter, but even the former is rather ductile compared with Narmco 5208, for example. The four different fibers used are listed in the order of decreasing tensile failure strain. In the table, WY is a high-strain graphite fiber and P75S is a high-modulus pitch fiber, both manufactured by Union Carbide. The glass fiber bundle contained about 200 filaments whereas the graphite fiber bundles had about 3000 filaments each. The matrix properties were measured but the fiber properties were taken from manufacturers' data sheets.

MATRIX PROPERTIES

Epoxy	Modulus, GPa	Ultimate tensile stress, MPa	Tensile failure strain, %
Epon 828/Z (80/20)	3.45	85.4	9
Epon 815/V140 (60/40)	2.13	45.5	14

FIBER PROPERTIES

Fiber	Diameter, μm	Cross-sectional area of bundle, mm^2	Modulus, GPa	Tensile failure strain, %
E-Gl	13.5	2.9×10^{-2}	72.35	4.8
WY-0224	5.1	9.16×10^{-2}	230.00	1.83
T300	7.0	11.61×10^{-2}	234.00	1.34
P75S	9.7	14.84×10^{-2}	517.00	0.40

COMPRESSIVE FAILURE STRAINS OF FIBER BUNDLES

E-glass fiber bundle has the highest compressive failure strain while P75S has the lowest. The high-strain graphite fiber is slightly stronger than T300. As expected from the buckling theory, the stiffer epoxy yields higher failure strains. However, the difference disappears for the high-modulus graphite P75S. The reason is that P75S fibers under compression fail in shear while the other fibers fail in buckling. Failure of P75S fiber was quite difficult to detect because the filaments failed individually on a plane about 45° to the loading. The failure was quite gradual, spreading over the entire length of the filament. Therefore, there was no sudden release of energy as in the other fibers. The data for P75S fiber should be taken as an estimate based on the best ability to detect failure. Note also that the strains were calculated from the failure stresses under the assumption of linear behavior.

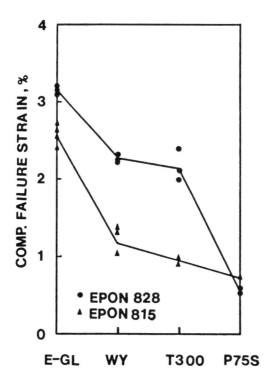

A CORRELATION BETWEEN COMPRESSION AND TENSION FAILURE STRAINS

Compressive failure strain increases with tensile failure strain. P75S graphite fiber has almost equal strength in both tension and compression. In the absence of buckling, T300 and WY fibers will be stronger in compression than in tension. E-glass fiber bundle buckles in compression before reaching a strain equal in magnitude to its tensile failure strain. Both T300 and WY fibers have almost the same modulus. Yet, T300 fiber is slightly weaker than WY fiber although the former is larger in diameter. Thus the same mechanism that controls tensile failure seems to affect buckling. For example, the same defect may become detrimental in compression as well as in tension. However, the defect sensitivity in buckling would be less than in tension because the superiority of WY fiber over T300 fiber is more pronounced in tension than in compression.

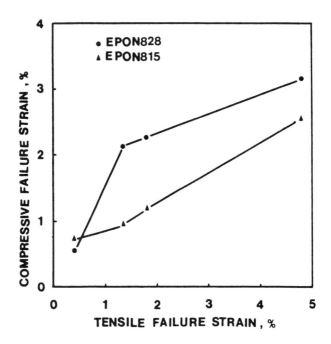

ANALYTICAL PREDICTIONS

Buckling of a fiber in an infinite matrix has been analyzed by several investigators [2,3]. An approximate equation for buckling strain can be derived by using the virtual work principle under the assumption that the matrix support of the fiber is proportional to the matrix modulus and the buckling pattern is sinusoidal. The predicted buckling strain is proportional to the square root of the matrix-to-fiber modulus ratio E_m/E_f. Further, the buckling wave length is inversely proportional to the fourth root of E_m/E_f. On the other hand, theories [4] on microbuckling of fibers in composites indicate that the buckling strain for extension mode, i.e., out-of-phase buckling, is proportional to the square root of the modulus ratio, as in the single-fiber case. However, the buckling strain for shear mode is predicted to be proportional to the modulus ratio itself. Thus, it is of interest to find out which theory can describe the experimental data obtained.

BUCKLING OF A FIBER IN MATRIX

- Sadowsky, Pu and Hussain, 1967 (ref. 2)
  Lanir and Fung, 1972 (ref. 3)
- Virtual work principle
- Matrix support proportional to modulus
- Buckling strain
  $\varepsilon \propto (E_m/E_f)^{\frac{1}{2}}$
- Buckling wave length
  $\delta \propto (E_m/E_f)^{-\frac{1}{4}}$

BUCKLING OF COMPOSITE

- Rosen, 1965 (ref. 4) and others
- Extension mode
  $\varepsilon \propto (E_m/E_f)^{\frac{1}{2}}$
- Shear mode
  $\varepsilon \propto E_m/E_f$

DEPENDENCE OF COMPRESSIVE FAILURE STRAIN ON MODULUS RATIO

Compressive failure strain is seen to be fairly proportional to the square root of the matrix-to-fiber modulus ratio. Even the high-modulus graphite fiber seems to follow the trend although its failure is not due to buckling. As will be shown in the following micrographs, the buckling of fiber bundles was in a characteristic shear mode. For microbuckling in composites, analysis predicts a linear relationship between the buckling strain and the shear modulus ratio. Therefore, the prediction equations for composites are not applicable to the present fiber bundles. However, the equation for buckling of a fiber in matrix correctly predicts the trend. It is thus possible that buckling of the bundle is triggered by buckling of a single fiber and hence the buckling strain is proportional to $(E_m/E_f)^{\frac{1}{2}}$.

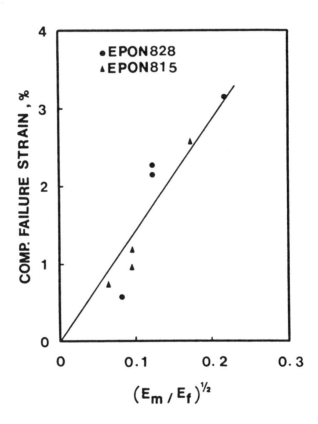

Composite Fracture Toughness and Impact Characterization

MODE OF BUCKLING FAILURE

A schematic diagram describing a bundle failed in buckling is shown below. Also shown is a shear failure of a fiber. The three parameters that deserve special attention are the segment length L, the buckling boundary angle α, and the orientation of fiber fracture surface. If buckling mode is sinusoidal, L is likely to be one fourth of wave length. Also, if fiber fails as a result of buckling, its fracture surface will be nomal to the fiber axis. Failure due to shear, on the other hand, will be associated with a slanted fracture surface. The buckling boundary angle, also called the crimp boundary angle, is generally less than 45°. All the foregoing parameters were examined on the failed specimens.

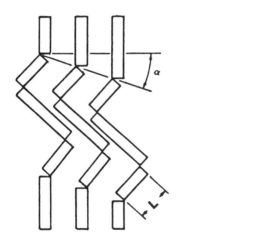

BUCKLING FAILURE   **SHEAR FAILURE**

BUCKLING FAILURE OF E-GLASS FIBER BUNDLE

The top figure shows buckling of E-glass bundle in a weak epoxy over its entire length. Too much solvent was inadvertently added to the epoxy during formulation and, as a result, was much more brittle and softer than expected. Buckling of the fiber bundle in this epoxy was quite gradual, starting at a very low load, and occurred uniformly over the entire length. The elliptical spots on the bundle are cracks growing into the matrix almost normal to the plane of the photograph. These cracks were generated by the tensile stress between fibers as a result of buckling. Note also that fibers are bent but have not broken yet. Buckling of the bundle in a well formulated Epon 815 us catastrophic and much more localized, as seen in the lower figure. Buckling occurs without warning and leads immediately to breakage of the fibers. Fiber break is seen to progress from one edge of the bundle to the other. The buckling failure of the E-glass bundle in the brittle Epon 828 is also quite localized, as seen in the figures on the next page. Although the bundle has an elliptical cross section, it can buckle in the plane of the larger dimension as well. The buckled region is out of focus because of the out-of-plane movement of the filaments. Macroscopically, no distinction could be detected between failures in Epon 815 and Epon 828.

UNIFORM BUCKLING AND MATRIX CRACKING, E-GLASS IN WEAK EPOXY

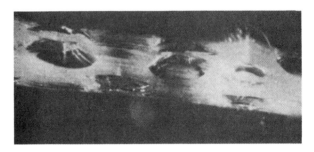

LOCAL BUCKLING OF E-GLASS BUNDLE IN EPON 815

LOCAL BUCKLING OF E-GLASS BUNDLE IN EPON 828: FRONT VIEW

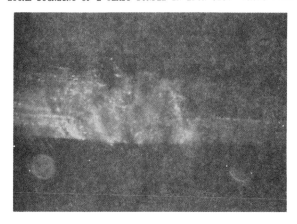

LOCAL BUCKLING OF E-GLASS BUNDLE IN EPON 828: SIDE VIEW

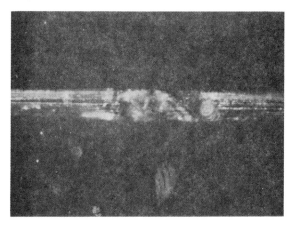

82 Tough Composite Materials

BUCKLING FAILURE OF GRAPHITE FIBER BUNDLES

The buckling failure of the T300 graphite fiber bundle in Epon 815 is similar to that of the E-glass fiber bundle in Epon 828. However, the buckling zone of the T300 bundle in Epon 828 is much narrower. Also, there is very little indication of buckling failure for the bundle in Epon 828. Rather, only a thin dark line indicates failure of the bundle. The high-strain WY graphite fiber behaves similarly to T300: the buckling failure is quite localized and the buckling zone in Epon 282 is smaller than in Epon 815.

LOCAL BUCKLING OF T300 GRAPHITE BUNDLE IN EPON 815: FRONT VIEW

LOCAL BUCKLING OF T300 GRAPHITE BUNDLE IN EPON 815: SIDE VIEW

Composite Fracture Toughness and Impact Characterization 83

LOCAL BUCKLING OF T300 GRAPHITE BUNDLE IN EPON 828: FRONT VIEW

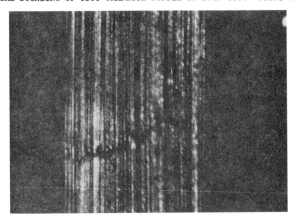

LOCAL BUCKLING OF T300 GRAPHITE BUNDLE IN EPON 828: SIDE VIEW

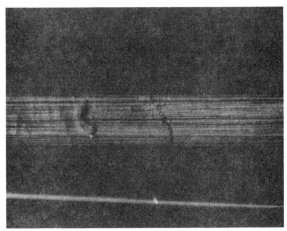

SEM MICROGRAPHS OF BUNDLES

After compression tests, one specimen from each group was cracked open through the bundle. The scanning electron micrograph of the E-glass bundle clearly shows a buckling-induced failure. The buckling is over the distance of several wavelengths. The rotation of broken fiber segments is clearly seen. The failure mode in Epon 815 was similar to the one shown below. The T300 graphite bundle shows more localized buckling failure than the E-glass bundle, as seen in the upper figure on the next page. As in tension, buckling will occur first at the weakest point. If fiber failure does not follow immediately, buckling may spread along the fiber axis. Since graphite fibers have lower tensile failure strain than glass fibers, the former are more likely to fail immediately after buckling. This may explain why fewer waves are involved in the buckling failure of graphite bundles. The failure mode of WY fiber observed in SEM micrographs was similar to that of T300 fiber. Note in the figures that the filaments on the left side of the fiber breaks remain bent. The high-modulus P75S fiber failed in shear without buckling, as seen in the lower figure on the next page. Even after failure the filaments remained straight without rotation or curvature. The figure shows a slanted fracture surface. Whereas the buckling failure of the other fibers was at a few isolated sites along the fiber length, the shear failure was quite uniformly distributed over the entire length. In other words, failure of P75S fiber was more like the multiple fracture frequently observed in tension of a fiber embedded in epoxy.

SCANNING ELECTRON MICROGRAPH OF E-GLASS BUNDLE IN EPON 828

SCANNING ELECTRON MICROGRAPH OF T300 GRAPHITE BUNDLE IN EPON 815

SCANNING ELECTRON MICROGRAPH OF P75S GRAPHITE BUNDLE IN EPON 828

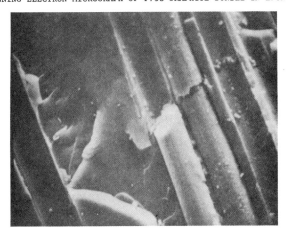

GEOMETRIC DESCRIPTION OF BUNDLE FAILURE

The number of fiber breaks and segment lengths in a typical failure zone is shown in the figure. The geometrical details of the failure zone vary much more than could be described by average numbers alone. Yet, the data in the figure indicate a certain trend. As was seen earlier, the failure zone of P75S fiber is a single, slanted fracture surface. Failure of WY fiber in both epoxies and T300 fiber in Epon 828 is characterized by double breaks in each filament. The number of breaks increases to three for T300 in Epon 815. E-glass bundles exhibit more than three breaks in each filament regardless of the epoxy type. The length of each broken segment also increases with the number of breaks. Discounting the direct role of the internal structure of fiber, we can infer from the figure that an increase in the number of breaks as well as in the segment length is associated with a lower modulus and higher compression failure strain.

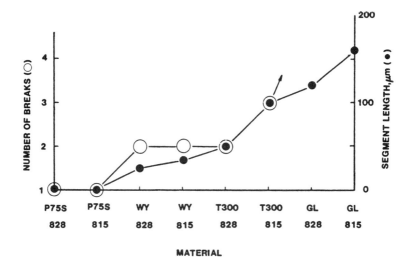

CONCLUSIONS

The use of a fiber bundle embedded in matrix can provide much needed information on compression failure mechanisms because failure of the bundle is well contained and can be monitored during testing. The method can clearly distinguish between buckling-induced failure and shear-induced failure. The present study indicates that WY and T300 graphite fibers and E-glass fiber fail in buckling while the high-modulus P75S graphite fiber fails in shear. Buckling-induced failure becomes more evident with low-modulus fiber in softer epoxy. Further specific conclusions are described below.

- Uniform buckling and matrix cracking in weak epoxy
- Buckling in plane of fibers
- Localized buckling in stiffer epoxy
- Higher buckling strain for E-gl than for graphite fibers
- Debonding after buckling
- Buckling strain increasing with tensile failure strain
- Fiber break at buckling except in weak epoxy
- Buckling strain proportional to $(E_m/E_f)^{1/2}$
- P75S fiber - No buckling, gradual compression failure, no buckling even after compression failure
- Failure more violent at higher strain
- Number of breaks in each filament highest for E-gl and lowest for high-modulus graphite
- Segment length longest in E-gl/Epon 815, and longer in Epon 828 than in Epon 815

REFERENCES

1. Williams, J. G. and Rhodes, M. D.: The Effect of Resin on the Impact Damage Tolerance of Graphite-Epoxy Laminates. NASA TM-83213, October 1981.

2. Sadowsky, M. A., Pu, S. L. and Hussain, M. A.: Buckling of Microfibers. J. Appl. Mech., Vol. 34, 1967, pp. 1011-1016.

3. Lanir, Y. and Fung, Y. C. B.: Fiber Composite Columns under Compression. J. Composite Materials, Vol. 6, 1972, pp. 387-401.

4. Rosen, B. W.: Mechanics of Composite Strengthening. Fibre Composite Materials, American Society for Metals, 1965, pp. 37-75.

THE EFFECT OF MATRIX AND FIBER PROPERTIES ON IMPACT RESISTANCE

Wolf Elber

*NASA Langley Research Center
Hampton, Virginia*

INTRODUCTION

Studies of impact damage in composites are aimed partly at evaluating the performance of structures with impact damage, and partly at establishing which basic material properties affect the impact resistance. Impact studies can be divided into three areas: impact dynamics, damage mechanics, and effects studies. Many of the studies in industry are in the last area. They seek to find residual compressive strengths after impact.

In our research effort we are concentrating on the other areas: impact dynamics and damage mechanics. The purpose of these studies (fig. 1) is to improve testing techniques and to identify the basic material properties which control impact damage. As a result, we expect to contribute to the criteria for selection or development of more impact-resistant matrix resins.

OBJECTIVES

* *Isolate basic material properties dominating the impact damage response of composite laminates.*

* *Show that matrix improvements alone cannot solve all problems (and that some tests are insensitive to matrix properties.)*

* *Show that fiber ultimate strain dominates the penetration phase.*

* *Show that for best impact resistance both fibers and matrix must be improved.*

* *Warn of the danger of unsafe failure modes for overly tough or strong matrix resins.*

Figure 1

FAILURE MODES IN THIN LAMINATES (I)

For thin laminates bonded over a circular aperture in a flat plate, the most severe damage from impact is usually observed on the back face. The specimen shown in figure 2 was a quasi-isotropic 8-ply plate of T300/5208. The low peel strength of the matrix resin results in extensive internal delamination and splitting, as well as peeling of the back face.

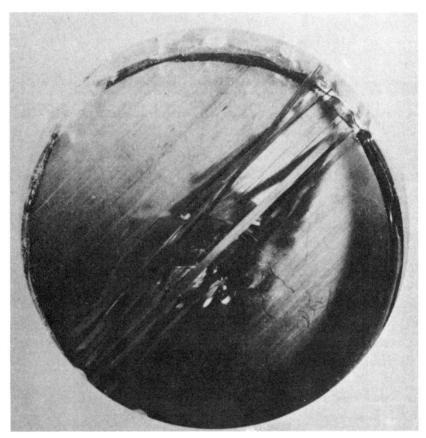

Figure 2

FAILURE MODES IN THIN LAMINATES (II)

For a tougher material (fig. 3) such as C6000/HX205, less peeling of the back face fibers is observed, and a cross-shaped transverse crack pattern accompanies the delamination.

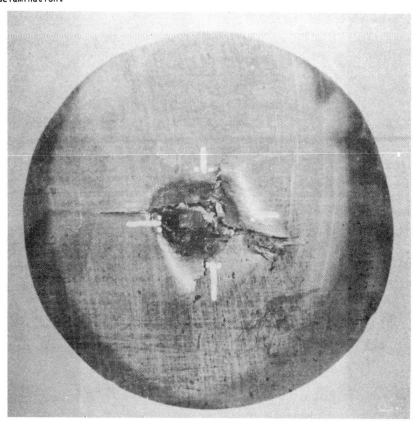

Figure 3

FAILURE MODES IN THIN LAMINATES (III)

For a thermoplastic resin such as polysulfone in C6000/P1700, the toughness and strength of the resin are so high that virtually no delamination occurs. Instead, a crossed pair of through-the-thickness cracks develops, allowing the impactor to penetrate the laminate by folding out the resulting leaves between the cracks. In the specimen shown in figure 4, one of these leaves broke during the penetration process.

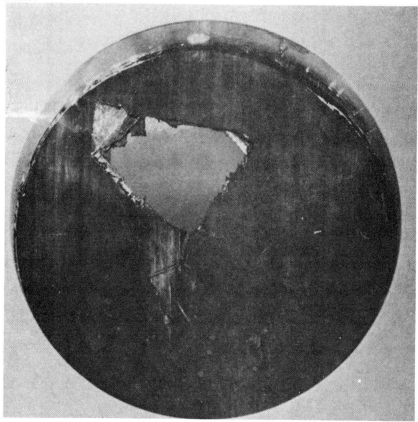

Figure 4

STATIC/IMPACT COMPARISON

Impact tests often result in local contact force histories including very high modes of vibration. Shown in figure 5 are the load-displacement relations for an impact test and for an equivalent static indentation test. In spite of the high noise spikes, the mean of the impact load-displacement curve follows approximately the curve for the static test. The curve for the static test, in turn, is bounded by the analytical curves for a plate and for a pure membrane in the region in which delamination occurs.

When both load-displacement curves are integrated to obtain energy absorbed as a function of displacement, the impact behavior is virtually identical to the static behavior (fig. 6). That implies that rate effects do not appear to affect the energy absorption and damage process, and also indicates that the high spike loads observed do not cause appropriate additional damage.

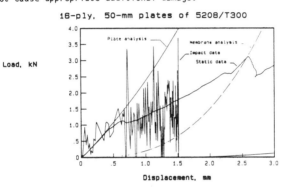

Figure 5

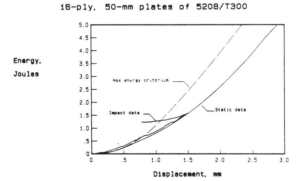

Figure 6

THIN PLATE ANALYSIS

Simple strength of materials formulations were found adequate to describe the deformation behavior of thin plates up to 32 plies (ref. 1). Figure 7 shows the superposition scheme for obtaining a total load displacement relation. First the flexural and shear displacements for a point-loaded plate are summed. For that summed midplane displacement, the large deformation membrane load term is obtained. The plate load and membrane reaction are summed to obtain the total load. The indentation displacement for the total load is added to the midplane displacement to calculate the total displacement. For large delaminations, the plate carries no load so that only the large deformation membrane load displacement relation is appropriate.

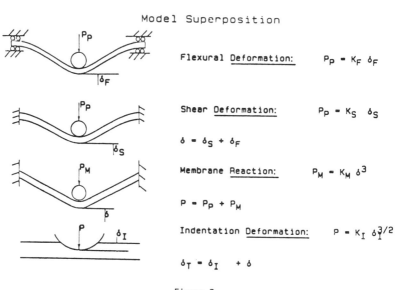

Model Superposition

Flexural Deformation: $P_P = K_F \, \delta_F$

Shear Deformation: $P_P = K_S \, \delta_S$

$\delta = \delta_S + \delta_F$

Membrane Reaction: $P_M = K_M \, \delta^3$

$P = P_P + P_M$

Indentation Deformation: $P = K_I \, \delta_I^{3/2}$

$\delta_T = \delta_I + \delta$

Figure 7

TYPICAL LOAD-DISPLACEMENT RELATIONS

The load displacement relation, and hence the impact resistance, is a strong function of the boundary conditions. Figure 8 shows the static load displacement relation for an 8-ply laminate over a 1/2-inch-diameter hole and over a 3-inch-diameter hole. In both cases the load displacement relation initially follows the analytical plate solution and approaches the fully delaminated or membrane solution when the damage is extensive. The small diameter plate shows a larger stiffness loss due to delamination, as would be expected.

two 8-ply circular plates

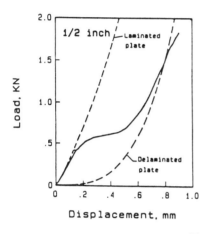
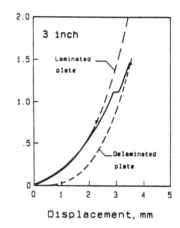

Figure 8

LOWER SURFACE FIBER STRAINS

The strength of materials models can also be used to calculate lower surface fiber strains, and figure 9 shows that the measured values of fiber strain basically agree with the calculations until splitting occurs in the lower ply.

clamped 2 inch diameter plate

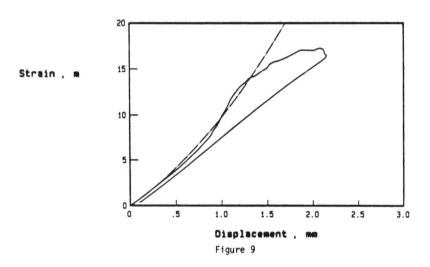

Figure 9

PLATE FAILURE CRITERIA

The strain expressions can be combined to calculate failure criteria as a function of plate size. If the plate is fully laminated, matrix shear failure must occur when the shear stress on the cylinder of the size of the indentation reaches a critical value. This is independent of plate size. As long as the plate is fully laminated, the strain in the bottom ply is a sum of the membrane and flexural strain. The middle curve in figure 10 describes the conditions when that strain reaches the fiber ultimate strain. But delamination starts at a lower load for this matrix (5208), so failure will occur when the membrane strain by itself reaches the fiber ultimate strain.

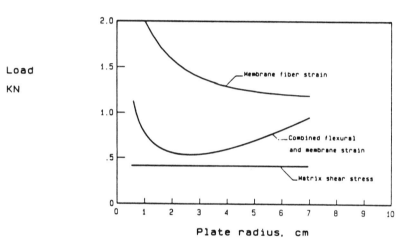

Figure 10

STATIC PENETRATION TESTS (I)

Figure 11 shows the load displacement relations obtained from static tests for 8-ply plates with three different support radii. The shear failure initiates at about 0.4 kN for all three plate sizes, but because these are thin plates, almost no stiffness is lost. In all cases the load displacement curves gradually approach the membrane state until fiber failures precipitate massive instabilities. In some cases the load may build up again until another ply (the bottom ply) fails.

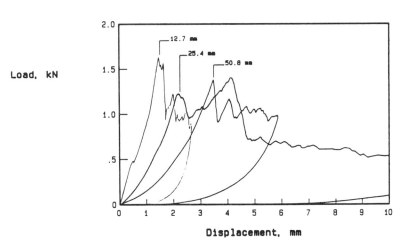

Figure 11

STATIC PENETRATION TESTS (II)

For the 16-ply plates shown in figure 12, shear stiffness is more significant and some stiffness loss is obvious at about 1.2 kN when the shear failures start in all three plate sizes. Again, the load displacement relations gradually become dominated by the membrane response and first-fiber failure is associated with significant stiffness loss.

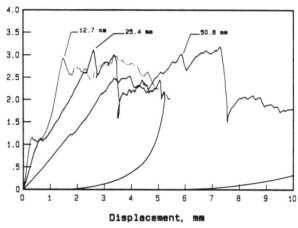

Figure 12

STATIC PENETRATION TESTS (III)

For the 32-ply data shown in figure 13, the shear failure instability is even more pronounced. Immediately after initiation, the delamination grows unstably to almost half of the plate diameter, with significant loss of stiffness. Further growth occurs as the displacement increases until the load displacement relation is almost completely membrane dominated and fiber failure begins in the lower plies.

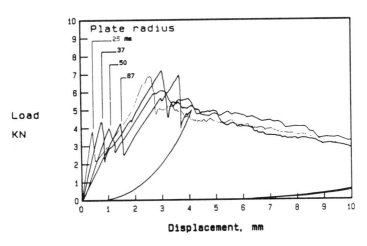

Figure 13

IMPACT FAILURE CRITERIA

The onset of delamination failure has been correlated with the punching shear stress on a cylinder of the diameter of the indentation circle. For the three thicknesses tested, a single critical shear strength describes the critical delamination load. This is shown in the left half of figure 14. The onset of membrane penetration by fiber failure has been correlated with the membrane tension strain under the load point. For the three thicknesses, a single value of ultimate strain described the critical displacements for both the first membrane instability point where one or more of the inner plies break and a second instability where the lower ply breaks. This is shown in the right half of figure 14.

5208/T300 GRAPHITE EPOXY LAMINATES

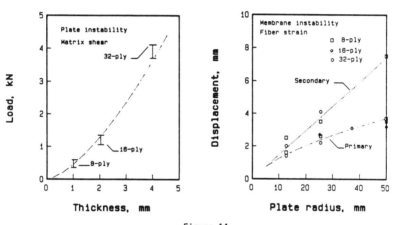

Figure 14

ENERGY LIMITS

Although generally the matrix properties are assumed to limit the impact resistance of composite laminates, improvements in matrix properties will result in fiber-initiated impact failures. In the left half of figure 15, the shaded area represents an upper bound on the impact energy up to the onset of penetration, based on the assumption that the matrix will not fail before fiber failure. In the right half of figure 15, this upper bound energy has been plotted against the support conditions for two types of fibers. The data from T300/5208 show that for the larger 8-ply plates, the impact resistance is essentially at the fiber limit.

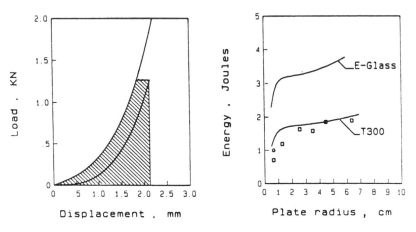

Figure 15

EFFECT OF MATRIX TOUGHNESS (I)

The polysulfone matrix P1700 is significantly tougher than the epoxy 5208. But, as expected, the load displacement plots in figure 16 for the two materials based on 8-ply 4-inch-diameter specimens do not show a difference because the failure is essentially fiber controlled.

8-ply, 100mm plates of 5208/T300 and P1700/C6000

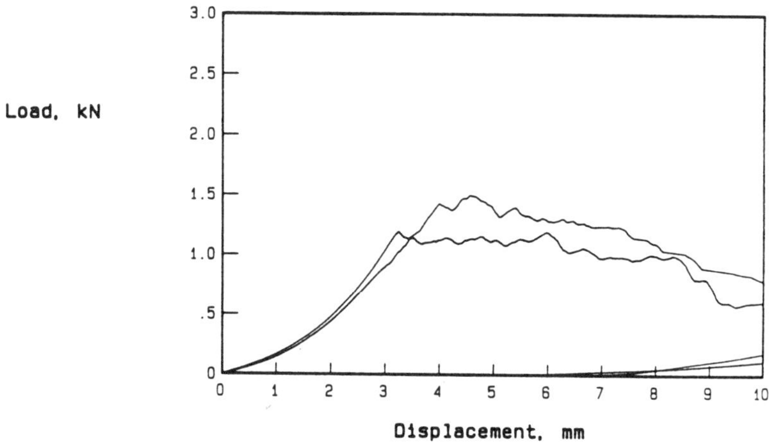

Figure 16

EFFECT OF MATRIX TOUGHNESS (II)

For smaller diameter plates, a small difference in load displacement relations in figure 17 is apparent. As in the 4-inch plate, fiber failure starts at smaller displacements than for the 5208 matrix because the stronger matrix retains much of the flexural stresses. The increased energy absorption comes largely during the penetration phase.

8-ply, 50-mm plates of 5208/T300 and P1700/C6000

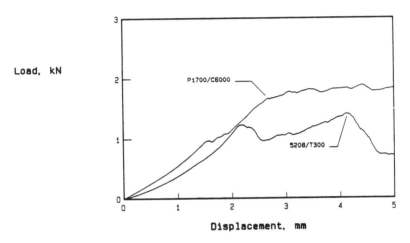

Figure 17

EFFECT OF MATRIX TOUGHNESS (III)

For the 1/2-inch-diameter support condition, the effect of matrix toughness becomes apparent before the onset of penetration, as shown in figure 18. Here the shear load on the matrix is high enough that matrix toughness contributes significantly to the energy absorption before the onset of fiber failure. In testing for impact resistance, it is therefore important to select the support conditions to reveal the effect of matrix properties.

8-ply, 13-mm plates of 5208/T300 and P1700/C6000

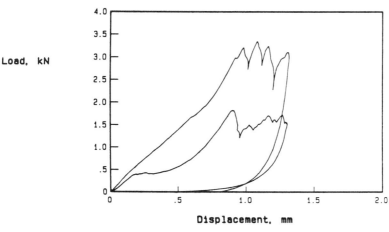

Figure 18

DELAMINATION ANALYSIS

The size of the delaminated area is a significant parameter in determining the residual compressive strength of impacted plates. Figure 19 shows that our static tests produce essentially the same delamination as impact tests taken to the same load or energy.

50 mm diameter 8-ply plates (T300/5208).

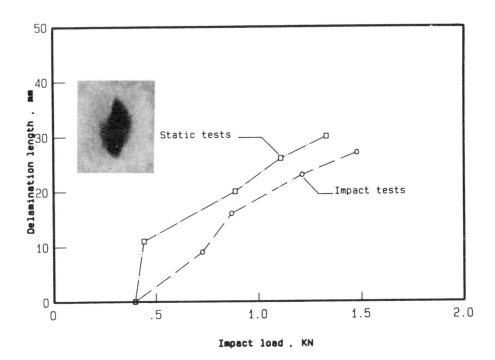

Figure 19

IMPACT FIBER DAMAGE

To reveal the amount of fiber damage and the sequence of damage, some impacted laminates were thermally deplied, separated into individual layers, and then divided into narrow fiber bundles representing essentially the original fiber tows. This process is shown in figure 20. These tows were then tension tested to establish the fiber damage from their residual strength (ref. 2).

FIBER BUNDLE TESTS
Specimen Preparation

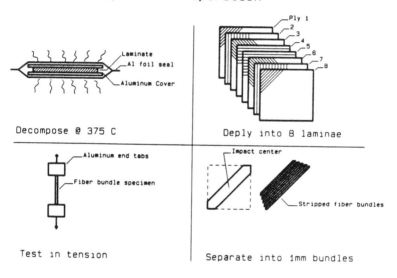

Figure 20

FIBER DAMAGE ANALYSIS

Figure 21 represents a scan of the residual tow strength in the vicinity of the impact center for an 8-ply plate tested to beyond the fiber damage onset. The most significant damage is found in the seventh ply (one layer from the back face). The eighth ply shows some damage, but no tow was completely broken. In the upper four plies (not shown), there was essentially no fiber damage.

Impact test to 1.04 Joules

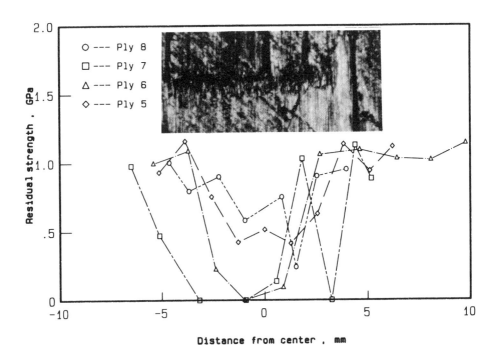

Figure 21

SUMMARY

Static indentation tests can provide better information about failure processes than impact tests, and they are simpler to conduct. In thin and thick laminates, we found that the matrix shear strength controls the onset of delamination damage and that the fiber strength controlled the penetration phase. In thin laminates, tough matrices do little to enhance impact resistance. Higher strain fibers would be required (fig. 22).

* *Matrix shear strength dominates the damage threshold especially in thick laminates.*

* *Matrix toughness dominates the type and extent of impact damage.*

* *Fiber ultimate strain dominates the membrane penetration energy.*

* *Togh composites require both tougher matrices and tougher fibers.*

* *Excessively strong matrix materials can result in brittle composites.*

Figure 22

REFERENCES

1. Bostaph, G. M.; and Elber, W.: Static Indentation Tests on Composite Plates for Impact Susceptibility. Proceedings of the Army Symposium on Solid Mechanics, 1982--Critical Mechanics Problems in Systems Design, AMMRC-MS-82-4, Sept. 1982, pp. 288-317.

2. Elber, W.: Failure Mechanics in Low-Velocity Impacts on Thin Composite Plates. NASA TP-2152, May 1983.

WORKING GROUP SUMMARY: COMPOSITE FRACTURE TOUGHNESS AND IMPACT CHARACTERIZATION

T.K. O'Brien, Chairman

The basic threat to the more widespread use of composites in aircraft primary structure is the problem of residual strength in the presence of damage. Specific problem areas that need to be addressed include the following:

1. Modeling/understanding composite failure
   a. Impact: relate strain energy release rate (G_{Ic}) to strain-to-failure
   b. Open hole compression: relate shear crippling to strain-to-failure
   c. Account for variables such as thickness and stacking sequence

2. Micromechanics models: develop a pragmatic model to serve as a "useful guide"

3. Consider the use of hybrid combinations, both interply and intraply, using graphite, Kevlar, and glass

4. Modeling/understanding the role of the interface

The chemists envision a strong need for an "all-inclusive" micromechanics model that will allow them to measure neat resin (and interface) properties and estimate their influence on composite properties. This need is especially strong in relating neat resin fracture toughness to composite interlaminar fracture toughness. Chemists continue to emphasize that specimen size should be "small."

The mechanics and material science community has many reservations about committing large amounts of dollars and resources to developing a single, all-inclusive micromechanics model. They point out that over the past 20 years micromechanics has been reasonably useful for predicting bulk properties (modulus) but has not been useful for predicting damage and ultimate failures (strength). Since toughness falls into the latter category, the mechanics community is very skeptical of the possibility of a successful micromechanics bridge of the gap between neat resin fracture toughness (and some interface property measurement) and composite interlaminar fracture toughness. The attempt to bridge this gap should be pragmatic. It would be best to identify those who have sound ideas for attempting to bridge the toughness gap and establish a small concentrated effort to develop, if not an all-inclusive model or strong predictive tool, at least a useful guide.

One concern about the development of micromechanics models to predict ultimate properties (toughness) was the fear that the major result would be a lot of solutions to boundary value problems. It was stated that what is needed instead is a multidisciplinary approach involving new physics, new concepts, and a lot of new material inputs to get a very realistic appraisal of how the constituents behave and how they interact together.

Concern was expressed that the bulk properties of matrix materials are not those that should be used in micromechanics models. The bulk properties we measure come from relatively large samples. In the composite, the bulk of the resin occupies 5 to 10 μm, and the resin properties at that scale may have only a limited relationship to those measured from large specimens. It was strongly suggested that this is a first-order effect and should be addressed soon, especially in view of the long-range program NASA has established.

Only the mechanics group focused on the gap between composite material and structural laminate properties. The group agreed that the most important and most urgent tasks were in this area. Although what needs to be done in this area is a little more clear cut, the problems are not simple or solvable in the near future. Further, solving these problems involves a multidisciplinary approach including dynamics, structural stability, composite mechanics, and fracture mechanics.

Part II
Constituent Properties and Interrelationships

The information in Part II is from *Tough Composite Materials,* compiled by Louis F. Vosteen, Norman J. Johnston, and Louis A. Teichman, NASA Langley Research Center, Hampton, VA, 1984. The report is the proceedings of a workshop sponsored by NASA Langley Research Center, May 1983.

FUNDAMENTAL STUDIES OF COMPOSITE TOUGHNESS

Kenneth J. Bowles

NASA Lewis Research Center
Cleveland, Ohio

INTRODUCTION

As a part of our work in the area of tougher composites at the Lewis Research Center, we are conducting fundamental studies of composite toughness. In this study, it is expected that relationships between neat resin matrix properties and composite toughness characteristics can be established. We can then use these relationships to either specify or evaluate candidate tougher composite matrices. It is also expected that the results from this study will contribute to the ongoing effort to better understand composite toughness.

The overall approach in this research program is presented in Figure 1. The first five steps have been completed and work to complete step six is now in progress. Two of these steps merit special discussion. The first is step one. Four resins with a wide range of mechanical properties were selected as matrix materials to highlight any subtle relationships which might exist between resin properties and composite toughness. In the selection process, special consideration was given to tensile modulus, tensile strength and strain to yield or failure. Step five, the evaluation of impact behavior and measurement of interlaminar fracture toughness, was the key step in which composite toughness was evaluated. These two tests were chosen to assess damage threshold and damage propagation characteristics of the composites. Only the impact test results have been analyzed and will be discussed in this presentation.

OBJECTIVE: ESTABLISH RELATIONSHIPS BETWEEN BULK RESIN PROPERTIES AND COMPOSITE TOUGHNESS

APPROACH:
1. SELECT FOUR RESINS WITH A WIDE RANGE OF MECHANICAL PROPERTIES.
2. CHARACTERIZE EACH OF THE FOUR RESINS AS COMPLETELY AS POSSIBLE.
3. FABRICATE COMPOSITES WITH THE SELECTED RESINS AS MATRICES.
4. MEASURE MECHANICAL PROPERTIES OF THE PROCESSED COMPOSITES.
5. EVALUATE IMPACT BEHAVIOR AND MEASURE INTERLAMINAR FRACTURE TOUGHNESS OF THE PROCESSED COMPOSITES.
6. DETERMINE VALID RELATIONSHIPS BETWEEN RESIN PROPERTIES AND COMPOSITE TOUGHNESS.

Figure 1

Constituent Properties and Interrelationships 115

MATERIALS

The materials that were used in this study are shown in Figure 2. The fiber is unsized Celion 6000 graphite fiber. The four resins are also listed along with general statements on the magnitude of the strength properties which were predetermined to be most important in this study. Three of these resins are epoxies and one is a polysulfone. The first two resins were used in the as-received state. The last two are mixtures of Ciba-Geigy 6010 and Ciba-Geigy 508, a flexibilized resin. The flexibilized 508 resin is a mixture of the 6010 resin and a polyol. Figure 3 presents the designation of the matrix resins, the type of polymer along with the applicable mixing ratio and the type of hardener for each of the four resins. In Figure 4 the general chemical structures are presented.

FIBER: CELION 6000.

RESINS:
- *FIBERITE 930 EPOXY. HIGH T.S. LOW STRAIN TO FAILURE. HIGH MODULUS.*
- *UNION CARBIDE P-1700 POLYSULFONE. HIGH T.S. HIGH STRAIN TO FAILURE. MEDIUM MODULUS.*
- *CIBA-GEIGY 6010-508-840 EPOXY. LOW T.S. HIGH STRAIN TO FAILURE. LOW MODULUS.*
- *CIBA-GEIGY 6010-508-956 EPOXY. LOW T.S. VERY HIGH STRAIN TO FAILURE. VERY LOW MODULUS.*

Figure 2

RESIN SYSTEMS

DESIGNATION	RESIN	HARDENER
930	DGEBA	DIAMINE
P1700	POLYSULFONE	
840	DGEBA:DGEBA + PO = 1:1	POLYAMIDE
956	DGEBA:DGEBA + PO = 2:3	TETRAMINE

Figure 3

RESINS AND HARDENERS STUDIED

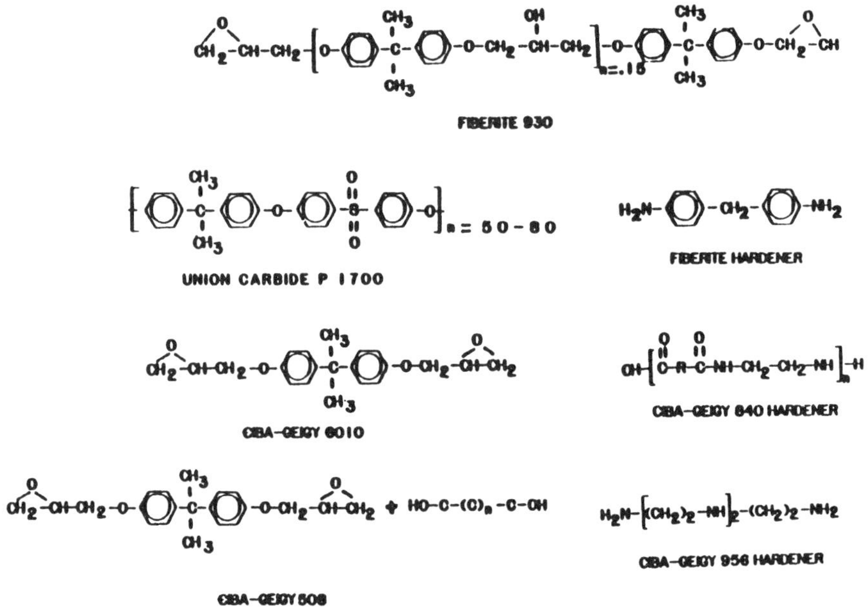

Figure 4

RESULTS

A listing of the more significant resin properties measured is shown in Figure 5. Most of the properties were measured using standard test methods except for the fracture toughness, impact and dynamic moduli. Fracture toughness testing was done using compact tensile specimens. The impact testing was done on six inch square plates, clamped along all four edges, with an instrumented drop-weight impact tester, and dynamic shear moduli were measured using a commercially available dynamic mechanical spectrometer. The resin mechanical and physical properties, with the exception of compression and relaxation data, are presented in Figure 6. The data shown in Figure 6 clearly show the wide range in tensile properties of the resins. Strain to failure was not measured for the P-1700 polysulfone because at a strain of about 10-15 percent, the specimens sustained severe necking in the gage section. The strain to failure was calculated from the reduction in area, but there was much scatter in the data.

The manufacturers mechanical properties data for the graphite fiber are presented in Figure 7. The values listed for Poisson's ratio and the longitudinal thermal expansion are best estimates provided by the vendor.

RESIN PROPERTIES

- *TENSILE PROPERTIES.*
- *COMPRESSIVE PROPERTIES.*
- *FRACTURE TOUGHNESS.*
- *INSTRUMENTED IMPACT.*
- *DYNAMIC MODULI.*
- *RELAXATION.*
- *DENSITY.*

Figure 5

BULK PROPERTIES OF MATRIX RESINS

RESIN	E(KSI)	σ_Y(KSI)	ϵ_Y	ϵ_F	ν	G(KSI)	SPECIF. GRAVITY	α(MM/MM°F) ++	T_G(°F) ++
930	657	12.9	--	0.019	0.36	236.4	1.335	29.5 x 10^{-6}	250
P-1700	344.6	11.9	.057	0.50+	0.37	134.6	1.240	31.5 x 10^{-6}	369
840	255	4.9	.03	0.35	0.44	82.6	1.146	64.7 x 10^{-6}	137
956	160	3.2	.05	0.50	0.365	52.2	1.179	38.8 x 10^{-6}	109

E = TENSILE MODULUS
G = SHEAR MODULUS
ν = POISSON'S RATIO
σ_Y = YIELD STRENGTH
ϵ_Y = YIELD STRAIN
ϵ_F = ULTIMATE ELONGATION

α = COEFFICIENT OF THERMAL EXPANSION
T_G = GLASS TRANSITION TEMPERATURE

+LOCAL NECKING PRECEDES FRACTURE
++TMA MEASUREMENT

Figure 6

CELION 6000 PROPERTIES

LOT	1231	2531
TENSILE MODULUS	34.4 MSI	34.7 MSI
TENSILE STRENGTH	513 KSI	533 KSI
ULTIMATE ELONGATION	1.49 %	1.54 %
POISSON'S RATIO	0.3	0.3
SPECIFIC GRAVITY	1.77	1.77
COEFF. OF THERMAL EXP. (LONG.)	-0.3 x 10^{-6}/°F	-0.3 x 10^{-6}/°F

Figure 7

NEAT RESIN IMPACT RESISTANCE

Figure 8 shows some of the results of the neat resin impact tests. The left ordinate is the amount of energy absorbed at failure. The right ordinate is the scale for the load at failure as measured by the instrumented tup of the tester. The ranking of toughness for the four resins is the same regardless of which of the two quantities is used as a basis for ranking (P-1700 >956 >840 >930). If the load at failure were chosen as the basis for comparison, the resin toughness would be a function of the strain at failure or yield. When one evaluates resin toughness on the basis of energy absorbed up to failure, one finds that the toughness is a function of the fraction of the load carried by the diaphragm action of the resin plate. This in turn is a function of yield strength and tensile modulus. From the results of this study, the toughness of neat resin materials, as assessed using the drop weight impact test, can be ranked using the resin stress-strain diagram.

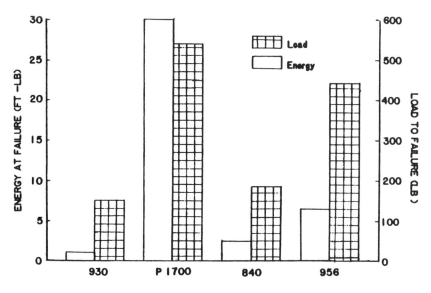

Figure 8

LOAD DEFLECTION CURVES

The instrumentation associated with the drop weight impact tester actually measures three variables. These variables are time, load and tup speed. The energy and displacement values are calculated from the last two of these measured values. In order to check the validity of these calculations, static load deflection tests were run on both resin and cross-plied composite plates. Figure 9 shows the results of these tests for a 956 resin plate. The solid line represents the load deflection curve from the impact tests. The open circles are data from the static tests. Disregarding the initial 0.08 inch deflection of the statically loaded specimen, the final portion of this curve is somewhat similar to the curve from the impact test. If the slopes of these two curves are considered as moduli, the modulus of the impacted specimen appears to be slightly greater than that of the statically loaded specimen after 0.08 inches of deflection. The same type of data are presented for a crossplied composite in Figure 10. There is very close agreement between the results of the two types of tests. From the results of this work, one can be assured that the data generated by the drop weight impact tester are valid data. Also, it is evident that plate displacement reactions to low velocity impact can be simulated by static testing methods.

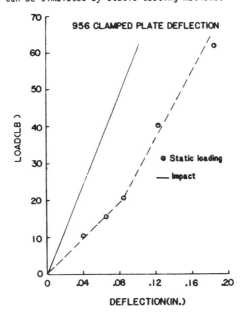

Figure 9

Constituent Properties and Interrelationships 121

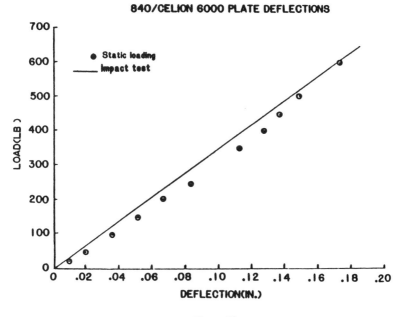

Figure 10

COMPOSITE PROPERTIES TESTS

Listed in Figure 11 are the tests that were used to generate the broad range of composite properties that are being studied for potential correlation with neat resin data. In this presentation only the results from the instrumented impact tests of crossplied composites will be discussed. The dynamic moduli and instrumented impact tests were conducted using the same equipment that was used for similar neat resin tests. The fracture toughness testing was done using double cantilever beam specimens. The impact tests utilized crossplied specimens having a thickness of 0.1 or 0.2 inches and unidirectional specimens having a thickness of 0.1 inch. Four specimens were tested for the 0.1 inch composites; two specimens were tested for the 0.2 inch composites; four specimens were tested for the unidirectional composites. The 0.2 inch thick specimens were subjected to an impact energy of 31 foot-pounds which was enough to initiate damage but not to cause thorough penetration. The 0.1 inch thick specimens were subjected to full penetration.

- *LONGITUDINAL TENSILE.*
- *TRANSVERSE TENSILE.*
- *10 DEGREE OFF AXIS.*
- *INTERLAMINAR FRACTURE TOUGHNESS (MODE 1).*
- *BENDING MODULUS.*
- *INSTRUMENTED IMPACT.*
- *DYNAMIC MODULI.*

Figure 11

COMPOSITE DROP WEIGHT IMPACT TEST TRACES

Load deflection traces are shown for each of the four types of thin, crossplied specimens in Figure 12. These traces show two different types of reaction to impact. The upper two traces for the 930 and P-1700 matrix composites show a linear increase in load to failure. After failure initiates, the load either drops off or maintains an approximately constant value. Traces of the 840 and 956 matrix specimens show a linear increase in load up to the point where some damage initiates. After damage initiation, the slope of the load deflection curve then decreases, but still remains linear up to the point of full failure where the load then drops abruptly. When one examines the energy curves (the lower curves in each plot), one can see that at full failure both the 930 composite and the P-1700 composite have absorbed only 6 foot-pounds of energy. In comparison, the 840 and 956 composites have absorbed 15 and 13 foot-pounds respectively. Any energy absorbed by the specimen after full failure is not useful energy. Visual examination of 840 and 956 composite specimens which were not fully penetrated indicated compressive failure at the point of incipient damage in the traces for the 840 and 956 composites in Figure 12. Even for an apparently fiber dominated test such as this, resin properties do appear to have an effect on the initiation of damage and possibly the propagation of damage through the composites.

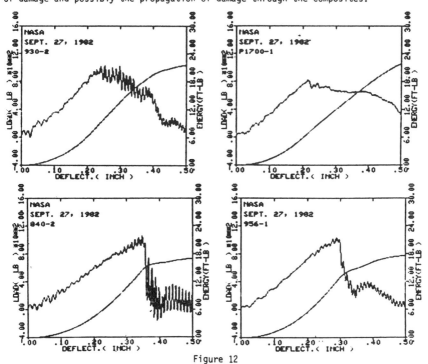

Figure 12

CONCLUDING REMARKS

Evaluation of drop weight impact testing of neat resin and crossplied graphite fiber reinforced composites has resulted in the development of three conclusions (Figure 13). Both geometric and composite constituent materials properties influence the drop weight impact resistance of crossplied composite plates. The matrix influence appears to be reflected in the incipient damage mechanism and the propagation of damage.

The future work in the area of tougher composites at the Lewis Research Center will be directed toward tougher, higher temperature composites. As a propulsion laboratory, we are interested in materials for aircraft engines. We propose to establish a data base for PMR-15 polyimide composites which are finding increased use in aircraft engine structures. This data base will be used as a reference against which we can evaluate the performance of newly developed, tougher, higher temperature composites.

1. DURING IMPACT OF CROSS-PLY LAMINATES, RESIN PROPERTIES APPEAR TO INFLUENCE THE EXTENT AND MECHANISM OF DAMAGE PROPOGATION.

2. BY ITSELF, NEAT RESIN STRAIN TO FRACTURE IS NOT A CONTROLLING INFLUENCE ON RESIN OR COMPOSITE FRACTURE TOUGHNESS.

3. IMPACT FAILURE ENERGY OF NEAT RESIN PLATES CAN GIVE A MISLEADING INDICATION OF THE RESIN CONTRIBUTION TO COMPOSITE IMPACT BEHAVIOR BECAUSE OF DIFFERENCES IN THE MAGNITUDE OF DIAPHRAGM ACTION IN THE TWO SYSTEMS.

Figure 13

INVESTIGATION OF TOUGHENED NEAT RESINS AND THEIR RELATIONS TO ADVANCED COMPOSITE MECHANICAL PROPERTIES

R.S. Zimmerman

University of Wyoming
Composite Materials Research Group

INTRODUCTION

Our program with NASA Langley at the University of Wyoming has been to evaluate four polymer matrix systems chosen for the ACEE technology program. We evaluated four epoxies, Hercules 3502, Hercules 2220-1, and 2220-3, and Ciba-Geigy Fibredux 914 epoxy, to judge their merits within the ACEE program and also to provide material property input for our micromechanics finite element program. During the first year we accomplished a number of tasks: we developed a casting procedure for the four matrix systems mentioned, generated mechanical properties at three test temperatures and two moisture contents and also performed a scanning electron microscope study to catalogue fracture surfaces for the unreinforced (neat) epoxies. We performed tension testing, shear testing, a limited amount of fracture toughness testing, and thermal expansion coefficient and moisture expansion coefficient testing. Mechanical material properties were generated for input into our micromechanics program and to characterize these four matrix systems.

TENSION SPECIMEN CONFIGURATION

The top specimen is the Hercules 3502 epoxy baseline system. This epoxy was chosen as the baseline since it is an untoughened epoxy. The second specimen is the Fibredux 914 epoxy from Ciba-Geigy. The third and fourth systems are the 2220-1 and 2220-3 epoxy systems. The specimens are 6 inches long and 1/10 inch thick. We used these dogbone specimens for all tension testing. They were fabricated using a steel box mold with metal dividers and cast as rectangular pieces. The specimens were then configured into the dogbone shape using a high speed router. (See fig. 1.)

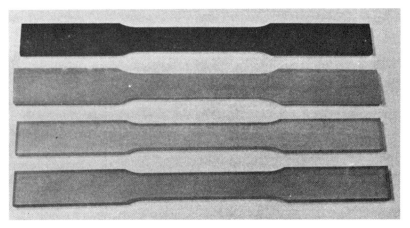

Figure 1

TORSION SPECIMEN

From top to bottom (fig. 2) the specimens are Hercules 3502 epoxy, Ciba-Geiby Fibredux 914, Hercules 2220-1 and 2220-3 epoxies. These were cast using two different methods. The three Hercules thermoset epoxies were easily cast using a steel mold. The Fibredux 914 epoxy from Ciba-Geigy was more difficult to cast. It had a much higher thermal expansion factor than the three systems from Hercules and was cast using a silicone rubber mold.

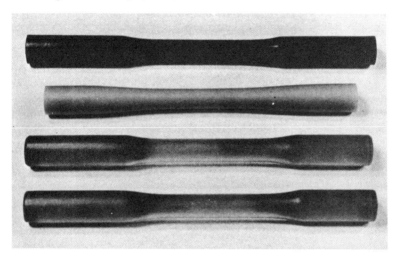

Figure 2

128   Tough Composite Materials

SQUARE STEEL MOLD WITH DIVIDERS AND SPACERS USED TO CAST FLAT TENSILE SPECIMENS

The steel mold in figure 3 was used to cast the 3502, 914 and 2220 epoxy systems. They were then machined into the dogbone shape, as seen in figure 1, using a high speed router.

Figure 3

STEEL MOLD AND RUBBER FUNNEL USED TO CAST ROUND DOGBONE TORSION SPECIMENS

The torsion specimens were cast into the dogbone shape directly, as shown in figure 4.

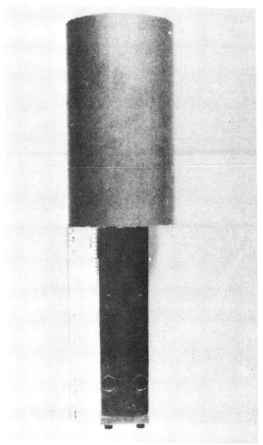

Figure 4

130 Tough Composite Materials

SILICONE RUBBER MOLD USED TO CAST THE CIBA-GEIGY FIBREDUX 914
IN THE ROUND DOGBONE SHAPE

The Fibredux 914 epoxy system was cast using the silicone rubber mold shown in figure 5 instead of the steel mold shown in figure 4 due to its slightly higher thermal expansion coefficient.

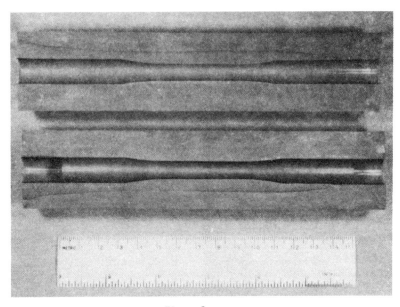

Figure 5

Constituent Properties and Interrelationships

TYPICAL TENSION FAILURE FOR FLAT DOGBONE SPECIMENS

The important features in neat resin tensile failures were noted in a vast majority of the failures. (See fig. 6.) A flat section, at A, is probably the initiation point of the failure. After initiation, a crack propagates across the specimen and then splits as it crosses the specimen and ejects a piece out of one side of the specimen, at B. These features are noted in all but a very few of our tests performed during this program. Note the two-axis strain gage at C.

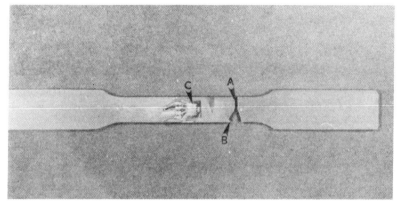

Figure 6

TYPICAL TENSION FAILURE SURFACE AS SEEN IN SCANNING ELECTRON MICROSCOPE (SEM)

We have characterized four specific areas in neat resin tensile failure. The first area is an initiation point, A. (See fig. 7.) It is typically a void or a piece of dust or some other impurity in the neat epoxy. Surrounding the initiation site is a relatively smooth area, B. The third area, C, is a transition between the smooth and a very rough area, D. Referring back to figure 6, the initiation site was located in the flat section and transitioned into the coarse area, where the triangular piece was ejected.

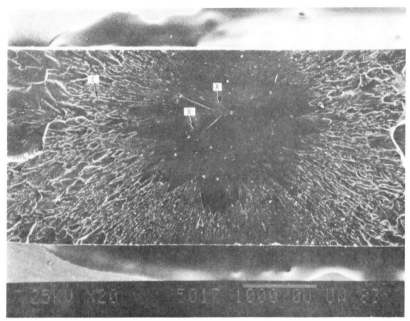

Figure 7

CLOSEUP OF TENSION FAILURE INITIATION SITE

This SEM closeup of figure 7 shows that this particular specimen had a small inclusion. One might note in this magnification that the inclusion itself is somewhere around 100 μm in size. (See fig. 8.)

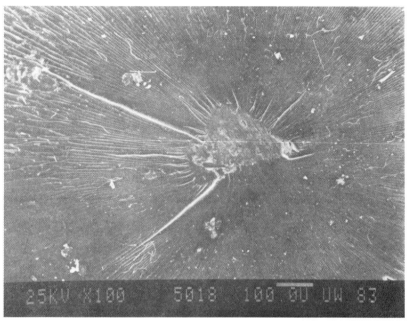

Figure 8

134 Tough Composite Materials

ATYPICAL FAILED TORSION SPECIMEN

Failures during shear testing can be demonstrated with this one-of-a-kind specimen (fig. 9). It failed at a relatively low strain compared to the other specimens, which helped preserve part of the failure process in the epoxy. The specimen is a 2220-1 tested at moisture saturation at 54°C. There is an internal 45° helix crack within this specimen. It is easy to visualize how the specimen basically unwinds and destroys itself as it fails. The vast majority of specimens shattered upon failure and were reduced to very small chips in the gage section. Although it is difficult to see in the figure, the feature to note is that the failure is at a 45° angle or on the tensile plane.

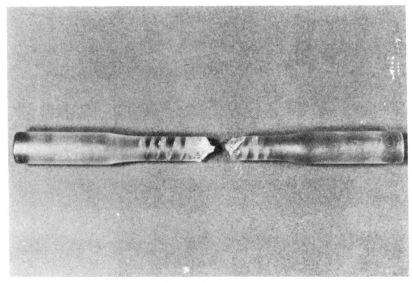

Figure 9

TORSION FAILURE SURFACE SHOWING AREAS OF FAILURE

In the torsion failure SEM photographs, we note the same four areas we saw in the tension failures. We have the initiation point either close to or on the surface as seen at A. It is surrounded by a fairly small smooth area at B with a transition area, C, into a rough area, D. (See fig. 10.)

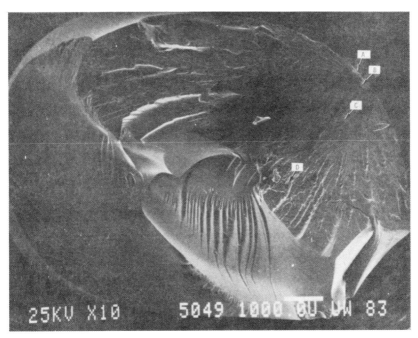

Figure 10

CLOSEUP OF TORSION FAILURE INITIATION SITE

This closeup of figure 10 shows the void just beneath the surface and affords a good view of the smooth area and the transition into the coarse areas of the failure. (See fig. 11.)

Figure 11

EPOXY STRESS-STRAIN DATA PLOTTED AT SIX DIFFERENT TEMPERATURE AND MOISTURE
CONDITIONS FOR HERCULES 3502, FIBREDUX 914, 2220-1, AND 2220-3

These data, taken on our computer data acquisition system, allowed us to plot the stress-strain curves to failure (fig. 12). These are actual data and these four figures are all plotted to the same scale to give a better idea of the representative strengths, strains-to-failure, and moduli of the four epoxies studied. Note on these four figures that the solid lines (dry test results) show a slight increase in strength with increasing temperature. They also show an increase in strain with increasing temperature and a slight decrease in modulus with increasing temperature. The dashed lines (moisture-saturated test results) appear in the same order (i.e., room temperature, 54°C, and then 82°C) and show a decrease in strength with increasing temperature and a decrease in modulus and a slight increase in strain values with increasing temperature. It should be noted that on the four systems studied, the moisture content was slightly different, with the 2220 systems showing lower moisture absorption. The moisture contents were as follows: 3.8% for 2220-1, 4% for 2220-3, about 5% for 3502, and 7% for 914.

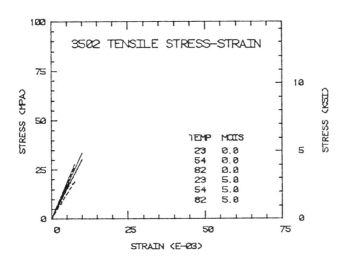

(a)

Figure 12

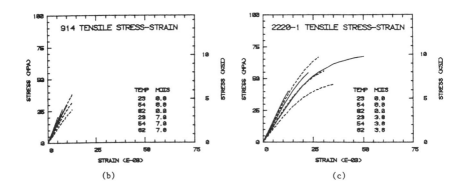

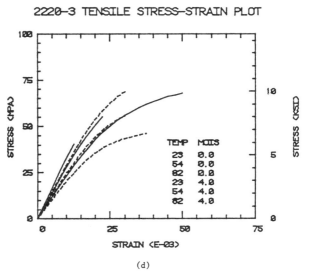

Figure 12.- Concluded.

AVERAGE VALUES FOR DRY AND MOISTURE-SATURATED TENSILE STRENGTHS FOR FOUR EPOXIES

Dry tensile strengths increase with test temperature, as shown in figure 13. The strengths for the 2220 systems are much greater than those for the 3502 or 914 systems in the dry state. Strengths decrease as temperature increases in all four moisture saturated resins.

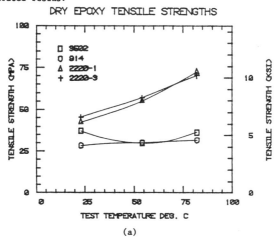

(a)

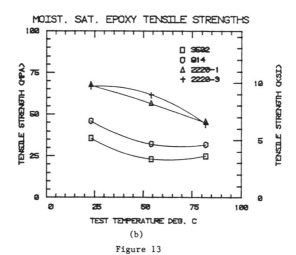

(b)

Figure 13

AS4/2220-1 GRAPHITE/EPOXY UNIDIRECTIONAL COMPOSITE LONGITUDINAL TENSILE STRESS-STRAIN RESPONSE AS PREDICTED USING MICROMECHANICS

Using the stress-strain curves of the neat resin for the matrix and an AS4 fiber for input, the longitudinal stress-strain response of the unidirectional graphite/epoxy composite was predicted using a micromechanics computer program developed at the University of Wyoming. (See fig. 14.) Note that the stress-strain response of the composite is highly linear to failure, as expected.

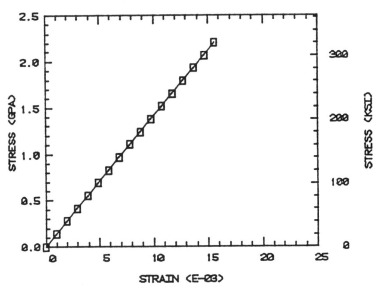

Figure 14

AS4/2220-1 GRAPHITE/EPOXY UNIDIRECTIONAL COMPOSITE TRANSVERSE
STRESS-STRAIN RESPONSE AS PREDICTED USING MICROMECHANICS

This micromechanics prediction plot starts out very linear but near the end there is
a slight amount of nonlinearity in the stress-strain curve. (See fig. 15.)

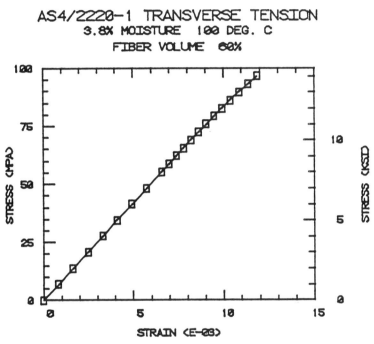

Figure 15

AS4/2220-1 GRAPHITE/EPOXY UNIDIRECTIONAL COMPOSITE LONGITUDINAL SHEAR STRESS-STRAIN RESPONSE AS PREDICTED USING MICROMECHANICS

Our micromechanics program has a nonlinear analysis capability. Figure 16 shows the shear stress-strain curve to have a high degree of nonlinearity.

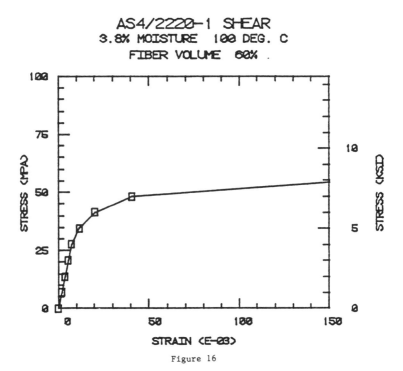

Figure 16

AS4/2220-1 GRAPHITE/EPOXY UNIDIRECTIONAL COMPOSITE, 100°C, 3.9 PERCENT
MOISTURE (ETW), 27.6 MPA (4 KSI) TRANSVERSE TENSILE APPLIED STRESS

A wealth of information is available from the micromechanical analysis. Figure 17 shows a typical example of the output. The grid assumes a square array of packing of the fibers. No stress contours are plotted within a fiber because they tend to clutter up the plot. The stress contours plotted in the matrix are octahedral shear, maximum principal stress. Almost any stress can be obtained from the analysis. Many input parameters can be varied, such as magnitude of applied stress, temperature, or moisture content in the matrix.

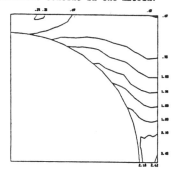

(a) Octahedral shear stress (ksi).

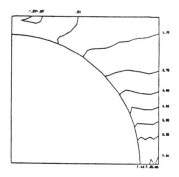

(b) Maximum principal stress (ksi).

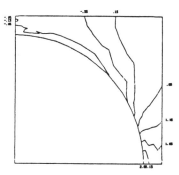

(c) Minimum principal stress (ksi).

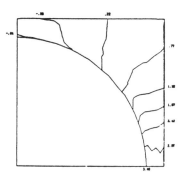

(d) Intermediate principal stress (ksi).

Figure 17

SUMMARY

Four neat resin systems were cast and their mechanical properties were generated. SEM photographs were analyzed and four specific areas common to all the specimens were identified on the fracture surfaces. The mechanical and physical properties were input into a micromechanics analysis and composite response was predicted for four different environmental conditions.

CONSTITUENT PROPERTY-COMPOSITE PROPERTY RELATIONSHIPS IN THERMOSET MATRICES

J. Diamant and R.J. Moulton

Hexcel Corporation
Dublin, California

Hexcel is committed to development of resins for aerospace structures and advanced composites. To execute this task effectively, a screening and evaluation methodology needs to be developed. Our procedure is depicted schematically in figure 1.

This entire procedure requires less than 300 grams of neat resin. Learning as much as possible from small amounts is especially important where long and costly synthesis routes are involved.

After a resin is formulated and compounded, it is necessary to establish the correct cure time-temperature relationship. The cure viscosity is conveniently measured with a Rheometrics Dynamic Spectrometer (RDS), thus establishing the gel time and temperature. Following cure, dynamic mechanical data of the cured resin are taken with the RDS. Thereby modulus (G') versus temperature behavior and T_g are established. Likewise it is easy to find out whether and/or how much postcure is needed above T_g. Loss mechanisms are apparent from the G" curve and in some cases may be related to mechanical and fracture behavior.

Tensile stress-strain measurements are performed on test specimens cast in dumbbell shape with a cylindrical cross section 3/16" in diameter and 1" long. Neat resin tensile modulus has been traditionally considered as a primary factor in matrix selection for primary composites (1).

High neat resin ultimate strain capability is needed for achieving high strain translation into composites, especially those made with high strain fibers. Laminate off-axis ultimate strain is reduced by a strain magnification of 5 to 10 times (2) and could be further complicated by different failure mechanisms in composites (3) compared to neat resins. Neat resin ultimate strain is one of several empirically known important parameters which has not been quantified. Compressive yield may be related to damage tolerance.

For G_{IC} measurements, 1 1/4" square coupons were cut from a 1/4" thick plate; compact double cantilever beam (CDCB) specimens were machined and tested according to ASTM E399 procedures (4).

Presently, mechanical tests are conducted under static conditions at 0.05 in/mi crosshead speed. Influence of testing rate, fatigue, and temperature will be investigated in the future. Instrumentation is being built now to investigate effects of moisture and measure accurately T_g at precisely known relative humidity. Neat resin physical properties, such as the coefficient of thermal expansion (CTE) and shrinkage in cure, are routinely measured for developmental resins. These properties may significantly affect mechanical and fracture behavior of composites, especially if the resin is brittle.

The purpose of the screening is to identify at an early stage matrices that will not have a chance of succeeding. But if a neat resin "passes," its laminate properties still may not be passable. A composite must be prepared at this step, and such variables as flow, fracture mechanisms, fiber-matrix interface, and translations of properties must be studied. Direct microscopic observation of load transfer from single fiber into matrix must be investigated. Longitudinal properties of unidirectional laminates are monitored and compared with predicted properties. This entire procedure serves as a semi-empirical method of scaling up the screening process from neat resin to composites.

146   Tough Composite Materials

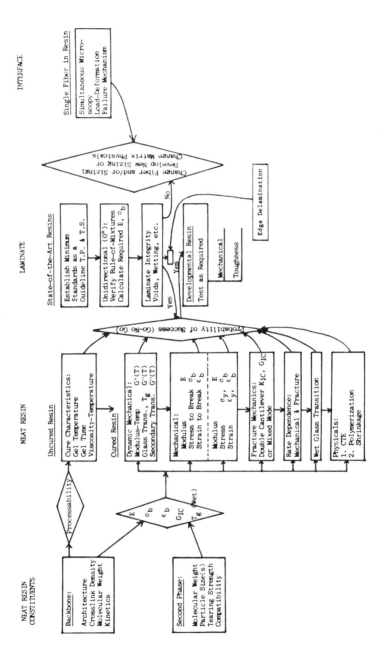

Figure 1. Resin screening and evaluation flowchart.

Constituent Properties and Interrelationships   147

The desired architecture of the polymer backbone and neat resin morphology are dictated by the major requirements of the finished composite. One of the principal current requirements is to improve damage tolerance without sacrificing service temperature and stiffness. On the polymer level, this means improved fracture toughness without lowering the wet modulus and wet T_g. One of the most accepted tests for damage tolerance is compression after impact. Evaluation of laminates with different Hexcel resins showed that the plot of compressive strength after impact versus the logarithm of G_{IC} followed a straight line (see fig. 2).

Hexcel resin F-185 has unacceptably low, dry, and wet T_g. On the other hand, F-263, a TGDDM/DDS type resin, complies with the stiffness and service temperature requirements, but because of its brittleness, its damage tolerance is unacceptably low. To comply with the Boeing standard, it appears that the G_{IC} of F-263 has to be increased by a factor of 40.

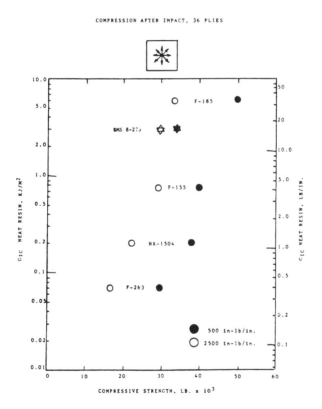

Figure 2. Neat resin G_{IC} versus laminate compression after impact.

The immediate question that arises is whether highly crosslinked resins can be made sufficiently tough. As a model, the TGDDM/DDS system was chosen because it is a state-of-the-art industry standard. To begin, toughening with carboxyl-terminated butadiene-nitrile (CTBN) rubbers was examined. Since toughening of highly crosslinked resins has not been thoroughly investigated, a more systematic view of the problem was taken. First the influence of rubber-resin compatibility was examined. Compatibility increases with increasing solubility parameter, which is directly related to the nitrile content. Second, the influence of rubber molecular weight was investigated. Third, the effects of rubber particle size were of interest. However, we do not have a way to control or regulate the rubber particle size, therefore it remains a dependent variable.

Figure 3 shows the chemical analysis, mechanical, fracture mechanics, and dynamic mechanical data of a TGDDM/DDS resin modified with 5% of different CTBNs. Chemical analysis of the pure CTBNs was performed. Titrations for epoxy and carboxyl were carried out and reported as equivalents per hundred (EPHR) and also converted to molecular weights using functionalities reported by the manufacturer. Molecular weights determined by GPC are approximately twice as high as those from titration. Below, $\bar{M}_n$ = 20,000 for all $\bar{M}_n$ values, but those measured by GPC agreed well. The GPC values were higher by a factor of 1.7. With increasing nitrile content a consistent increase in G_{IC} can be noted, while mechanical properties are unaffected. Low nitrile CTBNs produce G_{IC} values lower than those of the unmodified resin. Glass transition temperature is the same for all resins, including unmodified TGDDM/DDS, indicating good phase separation of the rubber.

	RESIN		DJ17 49-1	EB3-98A	EB3-98B	EB3-98D	EB4-28 Pre-reacted
TOUGHENING AGENT PROPERTIES	Toughening Agent(s) Type Toughener Wt. % Epoxy EPHR Carboxyl EPHR Titration $\bar{M}_n$ GPC $\bar{M}_n$ $\bar{M}_w$ Acrylonitrile % Solub. Parameter $\sqrt{cal/cm^3}$		None	CTBN 1300x31 5 - 0.052 3,654 6,792 13,556 10 8.45	CTBN 1300x8 5 - 0.054 3,333 7,419 15,220 18 8.77	CTBN 1300x13 5 - 0.060 3,000 6,571 12,864 26 9.14	CTBN 1300x13 5 - 0.060 3,000 6,571 12,864 26 9.14
MECHANICAL PROPERTIES	TENSILE	Modulus E msi Strength σ ksi Ult. Strain ε %	0.65 11 ≃ 2	0.56 8.4 1.75	0.59 9.1 1.40	0.56 8.2 1.3	0.56 9.6 4.5
	COMPRESSIVE	Modulus E msi Yield Stress σ_y ksi Yield Strain ε_y % Break. Stress σ_b ksi Break. Strain ε_b %	0.65 24 7 23 10	0.60 20.9 7.5 20.3 16	0.60 18.7 6 16.3 11	0.60 20.5 7 19.2 15	0.56 18.2 7 16 11.3
FRACTURE MECHANICS		Fract. Tough. K_{IC} ksi $\sqrt{in}$ Strain Energy Release Rate G_{IC} lb./in. G_{IC} KJ/m^2	0.58 0.45 0.079	0.33 0.17 0.030	0.40 0.24 0.042	0.53 0.44 0.078	0.57 0.51 0.089
DYNAMIC MECH.		Storage Mod. G' dyne/cm^2 Loss Mod. G'' dyne/cm^2 Glass Trans. Temp. T_g °C	1.70x10^{10} 3.95x10^8 243	1.48x10^{10} 3.30x10^8 242	1.53x10^{10} 4.09x10^8 242	1.48x10^{10} 5.31x10^8 242	1.46x10^{10} 4.04x10^8 243

Figure 3. The TGDDM/DDS resins modified with low $\bar{M}_n$ CTBNs (liquid rubbers).

Constituent Properties and Interrelationships   149

In figure 4 are shown TGDDM/DDS resins modified with medium and high $\bar{M}n$ rubbers. The former was synthesized by copolymerizing low $\bar{M}n$ CTBNs with a difunctional linear epoxy. Carboxyl titration indicates the extent of reaction whereas epoxy titration shows excess epoxy. The true $\bar{M}n$ may be regarded as approximately half the GPC value. G_{IC} values are higher than those obtained from epoxies containing the corresponding liquid rubbers, and increase with increasing nitrile content. Specimen DJ17-116D is an exception in this trend because of the lower degree of polymerization (low $\bar{M}n$) of the rubber.

High $\bar{M}n$ (solid) rubber Hycars 1472 and 1001CG are two very high $\bar{M}n$ butadiene-nitrile rubbers. Hycar 1472 has similar composition to 1300x13, but much higher $\bar{M}n$ and consequently numerous pendant carboxyl groups. Hycar 1001CG has no reactive groups. Because of the low solubility of these rubbers, GPC data reflect only the sol fraction. Both high $\bar{M}n$ rubbers give the highest G_{IC} values. However, the non-reactive Hycar 1001CG gives low tensile values, and so do the epoxy-CTBN copolymers, which have reduced reactivity due to greater chain length between the reactive ends. Hycar 1472 and prereacted CTBN 1300x13 give the highest tensile values. The conclusions from this series of experiments regarding toughening of TGDDM/DDS resins with elastomers are:
a. G_{IC} increases with increasing rubber $\bar{M}n$
b. G_{IC} increases with increasing CTBN nitrile content probably due to better compatibility with and adhesion to the matrix
c. Tensile ultimate properties are sensitive to the extent of reaction between rubber and matrix

Particle size is in the range of 0.5 to 1 micron for all $\bar{M}n$ values. Therefore, the $\bar{M}n$ effect is not due to difference in particle size. When rubber concentration is increased to 10%, a significant drop in modulus and T_g is registered. Therefore, addition of more than 5% rubber may not be desirable.

		RESIN	DJ17 48B	DJ17 84C	DJ17 84D	DJ17 116D	DJ17 117A	DJ17 117B
TOUGHENING AGENT PROPERTIES		Toughening Agent(s) Type	CTB 2000x162 & Epoxy	CTBN 1300x31 & Epoxy	CTBN 1300x8 & Epoxy	CTBN 1300x13 & Epoxy	Hycar 1472	Hycar 1001CG
		Toughener Wt. %	5	5	5	5	5	5
		Epoxy EPHR	.025	.0074	.017	.049	None	None
		Carboxyl EPHR	.0004	.0004	.007	.023	.06	None
		GPC $\bar{M}n \times 10^{-3}$	20	31	18	11	48	32
		$\bar{M}w \times 10^{-3}$	58	114	47	26	267	154
		Acrylonitrile %	0	10	18	26	27	≥ 27
		Solub. Parameter $\sqrt{cal/cm^3}$	8.04	8.45	8.77	9.14		
MECHANICAL PROPERTIES	TENSILE	Modulus E msi	0.55	0.54	0.58	0.49	0.57	0.50
		Strength σ ksi		7	7	7.5	13	4
		Ult. Strain ε %		3.5	2.5	2	5.5	2.5
	COMPRESSIVE	Modulus E msi	None	None	None	None	0.55	0.50
		Yield Stress σy ksi	"	"	"	"	21	15
		Yield Strain εy %	"	"	"	"	8	7
		Strength σ ksi	18.5	22	21	21	16	14
		Ult. Strain ε %	6	6.5	7.5	7.0	11	10
FRACTURE MECHANICS		Fract. Tough. K_{IC} ksi $\sqrt{in}$.	0.48	0.57	0.72	0.60	0.87	0.89
		Strain Energy lb./in.	0.36	0.53	0.78	0.64	1.16	1.39
		Release Rate G_{IC} KJ/m²	0.064	0.094	0.138	0.11	0.20	0.25
DYNAMIC MECHANICAL		Storage Mod. G' dyne/cm²	1.44x10¹⁰	1.40x10¹⁰	1.51x10¹⁰	1.28x10¹⁰	1.49x10¹⁰	1.28x10¹⁰
		Loss Mod. G'' dyne/cm²	3.81x10⁸	3.30x10⁸	4.07x10⁸	5.02x10⁸	9.64x10⁸	5.60x10⁸
		Glass Trans. Temp. T_g °C	244	245	244	242	240	243

Figure 4. The TGDDM/DDS resins modified with medium and high $\bar{M}n$ (solid) rubbers.

Toughening brittle plastics with rubber particles is the best known and oldest method, but not the only one. Another possibility is to use tough thermoplastics for modifying brittle thermosets. If G_{IC} can be significantly improved without loss of stiffness and T_g, then this could be a superior method. As a first try, ULTEM polyetherimide made by General Electric Company was used to verify the viability of this proposition (fig. 5). As shown, addition of 10% ULTEM does not significantly affect the dynamic shear and compressive moduli. Tensile data were scattered and therefore meaningful numbers could not be derived. The G_{IC} has doubled as compared to that observed for unmodified resin. Therefore, in principle, this method seems to work. The next question to address is to determine what thermoplastic is best to use and how much? Some additional insights and partial answers to these questions may be gained from dynamic mechanical data. However, before that subject is addressed, some additional possibilities for toughening are explored and compared in figure 5.

	TOUGHENING METHOD		LOW $\bar{M}$ ELASTOMER (Liquid Rubber)	HIGH $\bar{M}$ INERT ELASTOMER (Solid Rub.)	HIGH $\bar{M}$ REACTIVE ELASTOMER (Solid Rub.)	ABS COPOLYMER	TOUGH THERMO-PLASTIC	ULTRA-HIGH $\bar{M}$ GLASSY THERMOPL.	NONE
TOUGHENING AGENT PROPERTIES	Toughening Agent(s) Type		CTBN 1300x13	Hycar 1001CG	Hycar 1472	Blendex 311	Ultem	PEO	None
	Toughener	Wt. %	5	5	5	5	10	3	
	Carboxyl Titration	EPHR	0.060	None	0.060	None	None	None	-
	GPC	$\bar{M}_n \times 10^3$	3.0	-	-	-	-	-	-
		$\bar{M}_n \times 10^3$	6.6	32	48	-	-	5000	-
		$\bar{M}_w \times 10^3$	12.9	154	267	-	-	-	-
	Acrylonitrile	%	26	≥ 27	27	-	-	None	-
	Solub. Parameter	$\sqrt{cal/cm^3}$	9.14	-	-	-	-	-	-
MECHANICAL PROPERTIES TENSILE	Modulus	E msi	0.56	0.50	0.57				0.65
	Strength	σ ksi	9.6	4	13	5.6		10.5	11
	Ult. Strain	ε %	4.5	2.5	5.5	2.8		6	2
COMPRESSIVE	Modulus	E msi		0.50	0.55		0.63		0.65
	Yield Stress	σy ksi	18.2	15	21	19.5	20		24
	Yield Strain	εy %	7	7	8	8	5		7
	Break. Stress	σb ksi	16	14	16	18	19.5		23
	Break. Strain	εb %	11.3	10	11	14	7		10
FRACTURE MECHANICS	Fract. Tough.	K_{IC} ksi·$\sqrt{in.}$	0.57	0.89	0.87	0.62	1.09	0.57	0.58
	Strain Energy Release Rate	G_{IC} lb./in.	0.51	1.39	1.16		1.65		0.45
		G_{IC} KJ/m²	0.089	0.25	0.20		0.29		0.079
DYNAMIC MECH.	Storage Mod.	G' dyne/cm²	1.46x10¹⁰	1.28x10¹⁰	1.49x10¹⁰		1.65x10¹⁰		1.70x10¹⁰
	Loss Mod.	G'' dyne/cm²	4.04x10⁸	5.60x10⁸	9.64x10⁸		5.51x10⁸		3.95x10⁸
	Glass Trans. Temp.	T_g °C	243	243	240		190 & 244		243

Figure 5. Toughening of TGDDM/DDS with different modifiers.

An attempt was made to increase fracture toughness by addition of ABS copolymer, Blendex 311 (made by Borg-Warner), which is used routinely to toughen PVC. This material was found to be incompatible with the resin matrix and formed particles that could be seen with the optical microscope. There was no significant improvement in toughness and, therefore, mechanical testing was not completed. The addition of an ultrahigh $\bar{M}n$ modifier, in this case polyethylene oxide (PEO) with $\bar{M}n \approx 5 \times 10^6$, was also studied. It was found that the PEO was miscible with the cured matrix giving translucent specimens. But because of its high viscosity, only a small amount of modifier could be used. Again, G_{IC} has not improved, and elaborate mechanical testing was not justified.

The above experiments indicate that epoxy modification with glassy thermoplastics is a viable method which seems to be superior to elastomer toughening. In figure 6 mechanical, fracture toughness, and thermal data are shown of a number of commercial engineering thermoplastics which may be considered for toughening of thermoset matrices.

	Commercial Name Generic Type Manufacturer		LEXAN Poly- carbonate G.E.	UDEL P-1700 Poly- sulfone UCC	VICTREX 100P Poly- ether- sulfone ICI	RADEL-A Poly- aryl- sulfone UCC	PEEK APC-1 Poly- ether- ether- ketone ICI	ULTEM Poly- ether- imide G.E.
TENSILE	Modulus Yield Stress Yield Strain Break. Stress Break. Strain	E msi σ_y ksi ε_y % σ_b ksi ε_b %	0.345 9.00 7 9.50 110	0.36 10.20 5-6 20-100	 12.20 40-80	0.385 12.00 6.5 40.0 	0.56 13.20 10.15 150	0.43 15.20 7-8 60
COMPRESSIVE	Modulus Yield Stress Yield Strain Break. Stress Break. Strain	E msi σ_y ksi ε_y % σ_b ksi ε_b %	 12.50 	 40.00 				0.42 20.30
FLEX	Flex. Modulus	E msi	0.340	0.39	0.380	0.399		0.480
FRACTURE	Izod, Notched Fract. Tough. K_{IC} Strain Energy G_{IC} Release Rate	ft. lb/in. KJ/m ksi in. MPa m lb/in. KJ/m²	16.00 0.86 3.47±.46 3.17 32.45 5.67	1.30 0.070 3.07±.53 2.80 22.97 4.01	1.57 0.084 2.15±.13 1.96 10.70 1.87	1.60 0.085 3.28±.19 2.98 24.25 4.25	 6.63±.41 6.03 68.25 11.92	1.0 0.054 3.05±.20 2.78 18.98 2.21
THERMAL	Glass Trans. Temp. Melting Temp.	T_g °L T_m °L	145	195	203(HDT)	216(HDT)	143 332	217

Figure 6. Mechanical and fracture properties of some engineering thermoplastics.

Dynamic mechanical data shed additional light on the mechanisms of toughening. In figure 7 are shown dynamic storage and loss moduli, G' and G", respectively, of TGDDM/DDS modified with liquid CTBNs of different nitrile content. It can be seen that while G' is unaffected by nitrile content, G" increases over the whole temperature range as nitrile is increased. This is a result of better resin-rubber mixing (or increased interphase) (5,6) and compatibility as nitrile content is increased. Simultaneously, there is a slight decrease in T_g as compatibility improves.

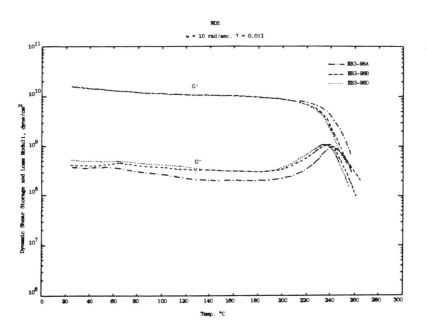

Figure 7. Dynamic moduli of TGDDM/DDS resins modified with low $\overline{M}n$ CTBNs containing 10% (-·-·), 18% (----), and 26% (·····) nitrile.

Figure 8 shows G' and G" curves of unmodified TGDDM/DDS and those modified with rubbers of different $\bar{M}n$ and reactivity but similar nitrile content. The former has the highest G' and lowest G", as expected. Highly reactive, high $\bar{M}n$ Hycar 1472 gives a lower G' because the rubber reduces stiffness. On the other hand, G" is much higher as a result of a large particle-matrix interphase resulting from the extensive reaction between each molecule of this rubber and the epoxy matrix. The low $\bar{M}n$ rubber produces a G' slightly lower than the previous material. However, G" is much lower due to smaller interphase. Apparently the reason is that each molecule has, at most, two reactive sites, but probably many of those are shielded by neighboring rubber molecules within the same particle, resulting in relatively low reactivity with the matrix. Still, G" is higher than that of the unmodified material, indicating some interphase. The nonreactive high $\bar{M}n$ rubber gives G' similar to the previous material, with G" slightly higher. This rubber has somewhat higher nitrile content and may interact with the matrix more extensively than the low $\bar{M}n$ rubber.

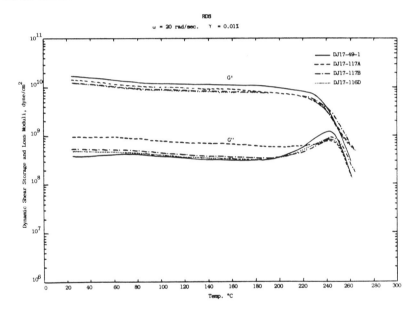

Figure 8. Dynamic moduli of TGDDM/DDS resins modified with different Hycar rubbers of high nitrile content: low $\bar{M}n$ CTBN (·····), high $\bar{M}n$ nonreactive rubber (-·-·), high $\bar{M}n$ highly reactive rubber (----), and unmodified resin (———).

Dynamic mechanical data of TGDDM/DDS resins modified with 10% ULTEM are shown in figure 9. It shows that the modified material has two peaks in G", corresponding to the unmodified resin and ULTEM glass transitions, respectively. The ULTEM T_g in the modified resin is shifted to a lower temperature, apparently due to plasticization. The two T_gs clearly indicate that this is a two-phase material. Again, G" is higher in the mixture than in either one of the ingredients, implying the presence of an interphase. Storage modulus, G', is not diminished by the presence of ULTEM.

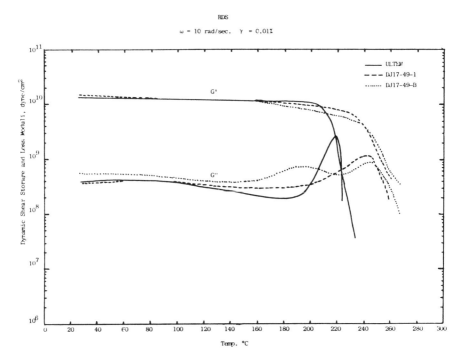

Figure 9. Dynamic moduli of TGDDM/DDS resin modified with 10% ULTEM (·····), unmodified resin (-----), and pure ULTEM (———).

CONCLUSIONS

The conclusions include:
1. Laminate damage tolerance correlates positively with neat resin G_{IC}
2. TGDDM/DDS resins can be toughened
3. Using butadiene-nitrile rubbers, neat resin G_{IC} and ultimate strain increase with increasing nitrile content, with the number of chemically reactive groups per molecule, and with increasing molecular weight
4. Better balance of properties combined with higher G_{IC} may be achieved by using tough glassy thermoplastics instead of rubbers to modify TGDDM/DDS resins
5. Both toughening mechanisms require good phase separation with an effective interphase
6. The kind and the amount of thermoplastic modifier remain to be selected
7. Presently only about a fourfold increase in G_{IC} was achieved, whereas a fortyfold increase is sought. Modification of the matrix backbone and of crosslinking is required

REFERENCES

1. Chamis, C. C., Hanson, M. P., and Serafini, T. T.: Modern Plastics, vol. 5, p. 90, 1973.

2. Christensen, R. M., and Wu, E. M.: Lawrence Livermore Laboratory Report UCRL-52169, November 1976.

3. Christensen, R. M., and Rinde, J. A.: Polymer Engrg. and Science, vol. 19, no. 7, May 1979.

4. ASTM E399 Standard. Standard Test Methods for Plane-Strain Fracture Toughness of Metallic Materials, 1981.

5. Diamant, J., Soong, D. S., and Williams, M. C.: Contemporary Topics in Polymer Science, vol. 4, 599-628, 1983.

6. Bates, F. S., Cohen, R. E., and Argon, A. S.: Macromolecules, vol. 16, no. 7, pp. 1108-1114, 1983.

THE EFFECT OF CROSS-LINK DENSITY ON THE TOUGHENING MECHANISM OF ELASTOMER-MODIFIED EPOXIES

R.A. Pearson and A.F. Yee

General Electric Company
Corporate Research and Development Center
Schenectady, New York

PART I: SUMMARY

In Part I of our NASA contract, a DGEBA epoxide resin (EPON 828 from Shell Co.) was elastomer-modified by using three different carboxyl terminated butadiene-acrylonitrile copolymers of varying acrylonitrile content (HYCAR CTBN elastomers from B.F. Goodrich Co.). The acrylonitrile content (10-27 wt%) dictated the size of the precipitated rubber particles in the cured epoxy matrix (under identical cure conditions). The curing agent, piperidine, was added in the amount of 5 parts per hundred parts resin (phr). The curing schedule consisted of 16 hours at 120 deg C. Materials containing 5 to 30 phr elastomer were investigated. The fracture toughness of these elastomer-modified epoxies was measured in terms of the critical strain energy release rate, G_{Ic}, using a three-point-bend geometry. The toughening mechanism was elucidated using a tensile dilatometry technique that directly measures the longitudinal and transverse strains, thus enabling the calculation of volume strain. A plot of volume strain versus longitudinal strain often readily reveals the types of micromechanical deformations occuring in the uniaxial tensile specimen up to yield. Several microscopy techniques were employed to corroborate the tensile dilatometry results (see Figure 1).

BACKGROUND

REVIEW OF LAST YEAR'S WORK:

  TOUGHNESS RESULTS– Toughness enhancement more a function of rubber content than rubber particle diameter

  ELUCIDATION OF TOUGHENING MECHANISM–
   Volume dilation of toughened materials indicate shear deformation & void growth
   Microscopy– shear & voiding deformations

  CONCLUSION– Rubber particles cavitate which dissipates bulk strain energy. The particles or cavities enhance shear band formation, which dissipates shear strain energy.

Figure 1

PART I: THE EFFECT OF RUBBER CONTENT

Epoxies composed of EPON 828/piperidine(5phr)/HYCAR CTBN 1300X8(varying content) are two phase materials consisting of an epoxy matrix with monodispersed one micron diameter rubber particles (abbreviated 828-pip-8). Epoxies composed of EPON 828/piperidine/HYCAR CTBN 1300X15(varying content) are two phase materials consisting of polydispersed 5-10 micron diameter rubber particles (abbreviated 828-pip-15). It was found that fracture toughness was more a function of rubber content than rubber particle size, for the 1-10 micron diameter range(see Figure 2). Also, it is important to note that Epoxy 828-pip-13 (13 is short for HYCAR CTBN 1300X13) did not contain a separate rubbery phase and exhibited little improvement in fracture toughness; thus the formation of a discrete rubbery phase seems to be essential to the rubber toughening of epoxies.

The addition of bisphenol "A" (BPA) to 828-pip-8 reduces the average particle diameter to ca. 5 microns at the 5 and 10 phr elastomer levels and induces a bimodal distribution of particle sizes (ten- and sub-micron) at 15 phr elastomer. Figure 2 also shows that the G_{Ic} values of the 828-BPA-pip-8 epoxies are greater than those of the 828-pip-8 epoxies. It is tempting to attribute the increased toughness enhancement to rubber particle size or size distribution. However, it will be shown later that this toughness enhancement is most likely due to the increased ductility of the epoxy matrix.

Note: it was observed that as the fracture toughness of these elastomer-modified epoxies increases, the size of the stress-whitened zone on the fracture surface of the 3PB specimens also increases; i.e., fracture toughness is directly related to the size of the stress-whitened or damaged zone.

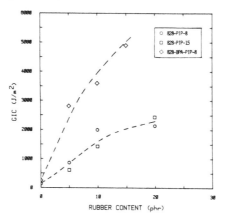

Figure 2

PART I: TENSILE DILATOMETRY DATA FOR BASELINE EPOXY

As mentioned previously, the toughening mechanism of these elastomer-modified epoxies was elucidated using tensile dilatometry. A servo-hydraulic testing machine was used to perform these experiments at several strain rates. Stress versus strain behavior as well as volume strain versus elongational strain behavior is monitored using this technique.

Figure 3 shows an engineering stress-strain curve and several volume strain curves taken at several strain rates for a neat epoxy. These curves are the results of an average of three tensile specimens tested for each strain rate reported. The stress-strain curve exhibits a yield stress and a reduction of stress after yield. This strain softening effect is indicative of shear band formation. Indeed, shear bands are easily detected with the naked eye on the surface of fractured tensile specimens. The intepretation of the volume strain curves is as follows: the initial increasing portion of the curve is simply due to the Poisson Effect; the curve then goes to a maximum volume strain, then to a slight decrease in volume strain which is indicative of shear deformation. Note that as the strain rate is increased the maximum in the volume strain increases and occurs at a larger elongational strain in a manner similar to that of the yield stress (not shown).

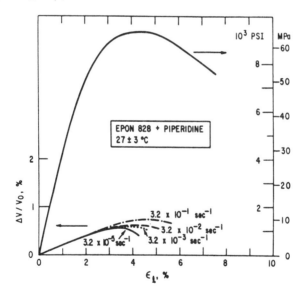

Figure 3

PART I: TENSILE DILATOMETRY DATA FOR ELASTOMER-MODIFIED EPOXY

Figure 4 contains the tensile dilatometry results of one of the elastomer-modified epoxies containing 20 phr HYCAR CTBN 1300X8. The set of stress-strain curves for this material clearly shows the increase of yield stress with increasing strain rate. At the lower strain rates the shape of the volume strain curves of these elastomer-modified epoxies is almost identical to that of the neat epoxy. However, at higher strain rates the volume strain curves are significantly different than those of the neat epoxy. At these high strain rates a process that causes an additional increase in volume strain occurs. This process is now known to be the voiding of the rubber particles. Also, a close examination of the volume strain curve taken at the highest strain rate reveals that enhanced shear deformation also occurs.

It is reasoned that these higher tensile strain rate results are more representative of the strain rates occurring at the crack tip. Thus, from the complete set of volume strain results (not all of the results are shown) the following toughening mechanism is proposed: the rubber particles, which are under a hydrostatic tensile stress at the crack tip, cavitate, which dissipates bulk strain energy. Furthermore, these cavitated particles promote the formation of shear bands which also dissipates bulk strain energy. It is the build-up of bulk strain energy that results in brittle fracture. In summary, cavitation and enhanced shear band formation are the micromechanical deformation mechanisms that occur ahead of the crack tip and produce a stress-whitened zone that effectively blunts the sharp crack, thus producing the fracture toughness enhancement.

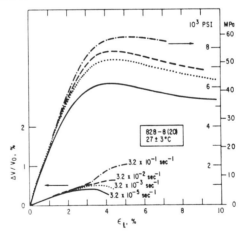

Figure 4

PART I: SCHEMATIC OF FRACTURE SPECIMENS

Scanning Electron Microscopy (SEM) and Transmission Optical Microscopy (OM) techniques were employed to corroborate the toughening mechanism elucidated using our uniaxial tensile dilatometry technique. Figure 5 is a schematic of the two types of fractured specimens used in this investigation. For brevity, let us concentrate on the stress-whitened region on the fracture surface of a SEN-3PB fracture toughness specimen, since the size of this region is related to fracture toughness.

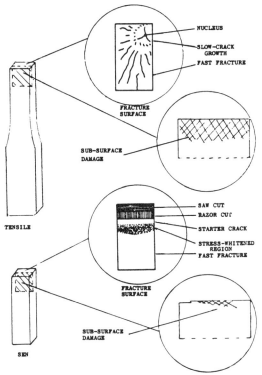

Figure 5

162   Tough Composite Materials

PART I: SEM MICROGRAPHS OF STRESS-WHITENED REGION AND FRACTURE SURFACE
         OF ELASTOMER-MODIFIED EPOXY

Figure 6 is an SEM micrograph of the stress-whitened region of 828-pip-15(30) that readily shows cavitation of the matrix and also signs of particle-particle interaction as evidenced by the non-hemispherical shape of the voids.

A SEM back-scattered electron imaging technique was employed on osmium tetroxide (OsO_4) stained fracture surfaces to prove conclusively that the "missing rubber particles" do not simply "pop out" but instead cohesively fail and are actually lining the walls of the cavitated matrix. Figure 7 is such a SEM micrograph. Note that the regions which contain OsO_4 appear white. This micrograph shows that the elastomer is in fact lining the walls of the cavities and the rubber particles are not "missing".

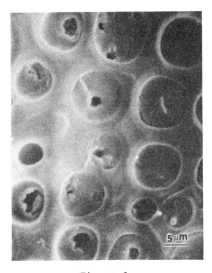
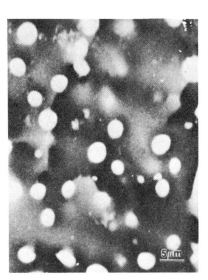

Figure 6   Figure 7

PART I: OPTICAL MICROGRAPH (OM) OF A FRACTURED ELASTOMER-MODIFIED EPOXY

OM is a useful technique for examining subsurface deformations which scatter light or induce birefringence. In order to view the damage below the fracture surface, a thin section perpendicular to the fracture surface was obtained by metallographic polishing techniques. Figure 8 is an optical micrograph of the subsurface damage in a SEN-3PB 828-pip-15(10) specimen which contains rubber particles greater than five microns in diameter, which are readily seen using this technique. In this micrograph, the thin section is viewed under bright field. The cavitated rubber particles appear dark since they scatter transmitted light. It is important to note that the oriented material composing the shear bands is more readily observed under crossed-polarized light.

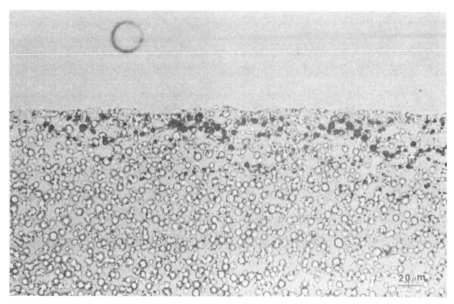

Figure 8

PART I: OPTICAL MICROGRAPH OF FRACTURED ELASTOMER-MODIFIED EPOXY UNDER CROSS-POLARIZED LIGHT

Figure 9 is an optical micrograph of the same thin section viewed between crossed polarizers. The birefringent regions which appear white are actually shear bands which connect the voided rubber particles. Shear banding and voided rubber particles are also seen in optical micrographs of thin sections of fractured tensile specimens (not shown).

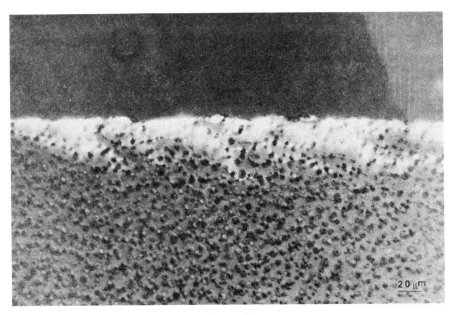

Figure 9

PART I: CONCLUSIONS

In summary, the types of deformations seen by the various microscopy techniques employed are completely consistent with our interpretation of the uniaxial tensile dilatometry data. It is important to note that Transmission Electron Microscopy (TEM) did not yield any evidence of even a single craze in these elastomer-modified epoxies and the tensile dilatometry results did not contain any evidence for the occurrence of a large diatational process; hence massive crazing as an energy dissipating or toughening mechanism is unlikely. We have presented evidence that these elastomer-modified epoxies are toughened by the formation of a plastic zone ahead of the crack tip. This plastic zone consists of cavitated rubber particles and shear bands whose formation effectively dissipates bulk strain energy thus inhibiting brittle fracture. Since most of the energy dissipation is due to the deformation of the epoxy matrix, then it seems reasonable to assume that enhanced matrix ductility should produce an even more pronounced toughening effect (see Figure 10).

PREDICTION:

Since most of the energy dissipation was due to matrix deformation then enhanced matrix ductility should produce an even more pronounced effect

Figure 10

PART II: OBJECTIVES AND APPROACH

OBJECTIVE:

The Part II objective was to determine the role of matrix ductility on the toughening mechanism of elastomer-modified epoxies.

APPROACH:

Our approach was to reduce cross-link density (hence increasing ductility) by using various equivalent-weight epoxide resins. Fracture toughness was again measured in terms of G_{Ic}. The characterization of the toughening mechanism was again performed using our uniaxial tensile dilatometry technique and corroborated using various microscopy techniques (see Figure 11).

2nd PHASE OBJECTIVE:

To determine the role of the matrix ductility on the toughenability and the toughening mechanism of elastomer-modified epoxies.

APPROACH:

Reduction of cross-link density by using epoxy resins of varying epoxide equivalent weights

Evaluation of fracture toughness by measuring Gic in 3PB

Characterization of toughening mechanism by tensile dilatometry and microscopy

Figure 11

Constituent Properties and Interrelationships   167

PART II: LIST OF MATERIALS

Figure 12 lists the materials used in this investigation. Several DGEBA epoxide resins where n=0 to 10 were used. These epoxide resin were cured with a stoichiometric amount of diaminodiphenyl sulfone (DDS). DDS was chosen due to its importance in aerospace applications. These epoxies were modified with a carboxyl terminated butadiene-acrylonitrile copolymer, a liquid elastomer commercially available from B.F. Goodrich (HYCAR CTBN 1300X13).

LIST OF MATERIALS

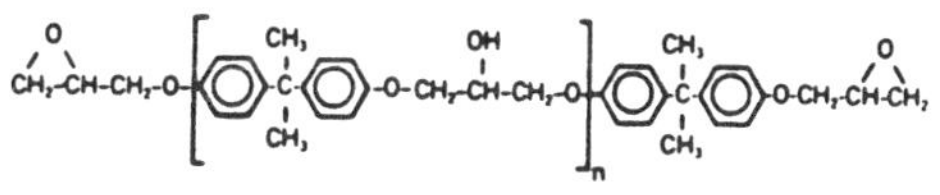

DGEBA Resin

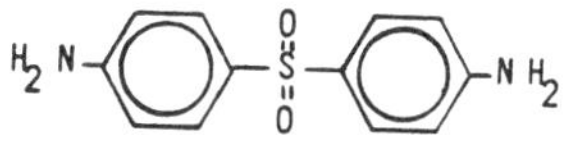

DDS (curing agent)

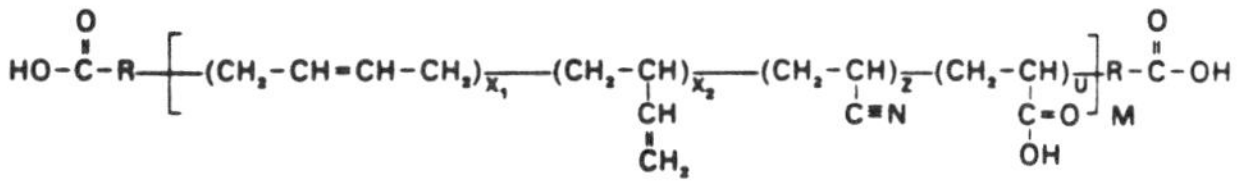

LIQUID ELASTOMER (HYCAR CTBN 1300x13)

Figure 12

PART II: FRACTURE TOUGHNESS RESULTS

Figure 13 is a plot of G_{Ic} versus epoxide monomer molecular weight. For the neat resins there is little dependence of G_{Ic} on monomer molecular weight. In fact all of these neat epoxies fail in a brittle manner when a sharp crack is present. As each of these epoxies is modified with 10% by volume elastomer an interesting effect is seen: the G_{Ic} becomes very dependent upon the epoxide monomer molecular weight. The highly cross-linked epoxies (low epoxide monomer molecular weight) are not toughened by the addition of the elastomeric phase. However, as the cross-link density is decreased the fracture toughness of the elastomer-modified epoxies increases; i.e., the toughenability increases.

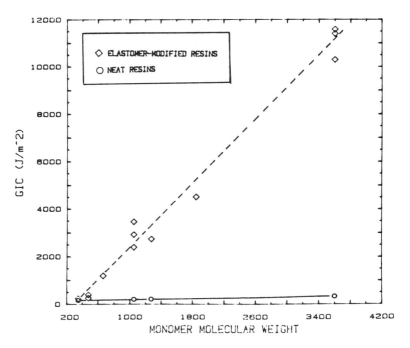

Figure 13

PART II: VOLUME STRAIN RESULTS

To elucidate the micromechanical mechanisms responsible for the enhanced toughness, volume strain measurements were again performed. However, only the highest strain rate was necessary. For the epoxy with the highest cross-link density, the volume strain data could not be measured since this material was very brittle even in uniaxial tension and fractured at less than 1% strain.

Figure 14 shows the tensile dilatometry results for an epoxy with a lower cross-link density. The solid lines are the results for the neat epoxy and the dot-dash lines correspond to the elastomer-modified epoxy. The reduction in the yield stress and tensile modulus upon elastomer modification is readily apparent in the stress strain curves. The volume strain curves exhibit the now familiar increase in volume strain due to the Poisson Effect. The slope of the volume strain curve decreases prior to yield, which indicates that a shear process has occurred. The slope of the volume strain curve of the elastomer-modified epoxy increases prior to yield, which indicates a voiding process has occurred.

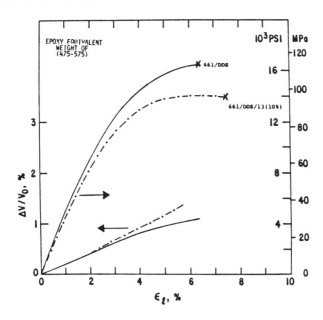

Figure 14

PART II: VOLUME STRAIN RESULTS

Figure 15 presents the uniaxial tensile data for the epoxy with the lowest cross-link density studied in this investigation. Note that in the stress-strain curves of both the neat and elastomer-modified epoxies there exists much extension after yield even at this high strain rate. These materials are very ductile compared to the epoxies with the highest cross-link density which are very brittle. The micromechanical deformation mechanisms are the same as in the other epoxies tested as evidenced by the volume strain curves. The neat resins tend to deform simply by shearing, while the elastomer-modified epoxies deform by voiding and shearing.

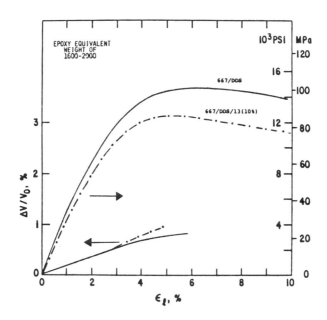

Figure 15

PART II: SEM ANALYSIS OF SEN-3PB FRACTURE SURFACES

SEM analysis of the fracture surface of the SEN-3PB specimens was again performed to corroborate our uniaxial tensile dilatometry results. Figure 16 is a SEM micrograph of the fracture surface of the elastomer-modified epoxy with the highest cross-link density. The crack had travelled from left to right and the area of interest is the small plastic zone which had formed prior to the propagation of the crack. Note that no stress-whitened zone can be detected with the naked eye since the small plastic zone was not sufficiently thick to scatter light. Also note that the shape of the fractured rubber particles are very hemispherical and are very shallow. The lack of shear deformation and cavitation of the matrix corresponds well to the lack of toughness enhancement seen.

Figure 17 is a SEM micrograph of the fracture surface of a SEN-3PB specimen of an elastomer-modified epoxy with a reduced cross-link density (monomer equivalent weight of 475-575 was used). For this specimen a stress-whitened zone is easily observed with the naked eye. The area of interest is again at the beginning of the plastic zone. The region before the plastic zone is of the precracked region and contains shallow cavities from fractured rubber particles. Note that in the plastic zone there exists evidence for localized shear deformation and the cavities appear to be much larger and deeper. The existence of the plastic zone corresponds well to the toughness enhancement.

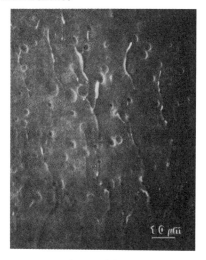 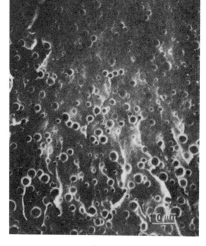

Figure 16   Figure 17

PART II: SEM ANALYSIS OF SEN-3PB FRACTURE SURFACES

Figure 18 is a SEM micrograph of the fracture surface of a SEN-3PB specimen of an elastomer-modified epoxy with the least cross-link density (monomer equivalent weight of 1600-2000 was used). The stress-whitened zone of this specimen spans the region where the starter crack had arrested to the edge of the specimen where the striker had made contact. This SEM micrograph shows that a stark contrast exists between the region of fast crack growth and the region of slow crack growth (the plastic zone). The plastic zone consists of rubber particles that have been sheared and cavitated to an extent much greater than in any of the other epoxies we have studied. The size of this plastic zone corresponds well to the increase in toughness enhancement.

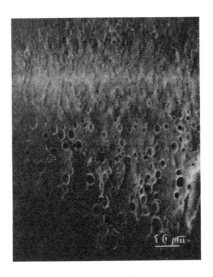

Figure 18

Constituent Properties and Interrelationships   173

PART II: OPTICAL MICROSCOPY OF SUBSURFACE DAMAGE

Optical Microscopy (OM) was used to observe the sub-surface microdeformations that had occurred in the SEN-3PB specimens.

Figure 19 is an optical micrograph of the plane perpendicular to the fracture surface (see figure 5) of a SEN-3PB specimen composed of an an elastomer-modified highly cross-linked epoxy. The thin specimen was viewed between crossed polarizers. There exists a shear zone of about three microns in thickness just below the fracture surface. Note that none of the particles appears dark; thus they did not cavitate. Also note the absence of shear bands. Since this plastic zone is relatively small, it is not surprising that no fracture toughness enhancement was seen.

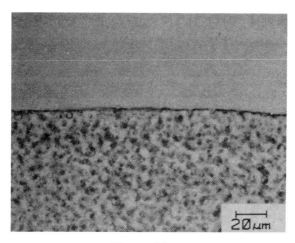

Figure 19

PART II: OPTICAL MICROSCOPY OF SUBSURFACE DAMAGE

However, as the cross-link density of the epoxy matrix is reduced for the elastomer-modified systems, the size of the stress-whitened zone and the value of G_{Ic} increase.

Figure 20 is an optical micrograph, again taken in polarized light and in the same plane as the previous micrograph. But in this specimen the cross-link density of the matrix was reduced by using an epoxide monomer of 475-575 g/eq-wt. The particles which appear dark have cavitated and thus scatter light. Close examination reveals that birefringent bands - shear bands - connect the cavitated rubber particles. It seems more than reasonable to attribute the enhanced energy dissipation to the formation of shear bands and voids which comprise the plastic zone.

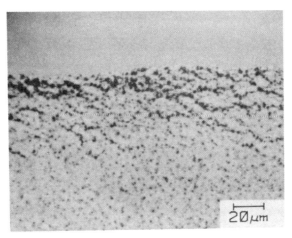

Figure 20

Constituent Properties and Interrelationships   175

PART II: OPTICAL MICROSCOPY OF SUBSURFACE DAMAGE

In this case figure 21 is an optical micrograph taken under identical conditons of a SEN-3PB in which the cross-link density was reduced by using an epoxide monomer of 1600-1800 g/eq-wt. The damage zone is much larger than those seen in the previously examined elastomer-modified epoxies. There is evidence for much cavitation and shear banding. As previously mentioned, the size of the plastic zone corresponds well to the increase in fracture toughness.

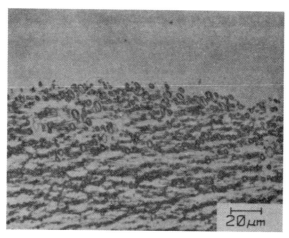

Figure 21

176 Tough Composite Materials

PART II: OPTICAL MICROSCOPY OF SUBSURFACE DAMAGE

Figure 22 is an optical micrograph of the same specimen but taken in the region where the specimen was subjected to a compressive stress due to the contact with the striker. This micrograph is important since it shows that these rubber particles can sustain large shear strains when subjected to a compressive stress.

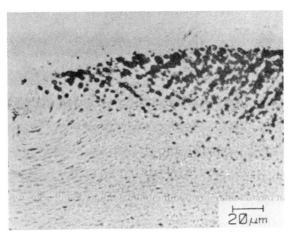

Figure 22

PART II: CONCLUSIONS

In conclusion, two points should be emphasized. First, the toughening mechanism of elastomer-modified epoxies consists of the formation of a plastic zone by voiding and shear banding. Second, increasing the ductility of the epoxy matrix enhances the toughenability of the matrix upon the addition of a second, rubbery phase. Therefore, as the cross-link density of the epoxy matrix decreases the fracture toughness of the elastomer-modified version increases. Figure 23 summarizes these conclusions.

CONCLUSIONS:

TOUGHENING MECHANISM– *When present, similar to piperidine cured systems => rubber particles cavitate and promote shear band formation*

EFFECT OF MATRIX DUCTILITY–

Lowering cross-link density enhances ductility

Increasing ductility enhances toughenability!

Figure 23

FREE VOLUME CONSIDERATIONS IN THERMOPLASTIC AND THERMOSETTING RESINS

Robert F. Landel, A. Gupta, J. Moacanin, D. Hong, F.D. Tsay,
S. Chen, S. Chung, R. Fedors, and M. Cigmecioglu

Jet Propulsion Laboratory
Pasadena, California

This paper describes some of the work which has been done at JPL on the direct measurement of physical ageing and of the volume changes which go on during the course of this ageing. I particularly want to stress the need for concomitant stress-strain and volume-strain measurements. We first present some data on PMMA showing its physical ageing as measured by its stress relaxation response and the accompanying volume changes; we follow with an indication of how one can measure the free volume both directly from the volume change itself and (we think) relatively directly via a new and independent technique; and finally we show the application of this new technique -- electron spin resonance spectroscopy -- to other polymer systems.

Figure 1 is a reminder of the course of stress relaxation in PMMA for samples which have been aged at room temperature for various lengths of time, after cooling from the molten state, before starting the experiment (Cizmecioglu et al.).

For each ageing time the modulus is higher, although (in these experiments) the rate of stress relaxation remains constant. Hence the data are superposable to a single master curve by a translation in the direction shown by the arrow.

This ageing process occurs because the material is in the glassy state and hence is not at volumetric equilibrium.

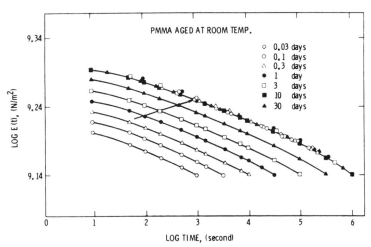

STRESS RELAXATION OF ANNEALED PMMA AT ROOM TEMPERATURE
CIZMECIOGLU ET AL.

Figure 1

To examine the consequences of not being at equilibrium, consider the volume-temperature-time response. On cooling a glass-forming liquid, the volume follows the liquidus line, shown in figure 2. At the glass temperature T_g, the expansion coefficient changes and the volume follows the path containing point A. If the cooling rate is reduced, the volume proceeds farther along the liquidus line to a lower T_g, and then follows the path containing point B. If the sample is taken to point A and held at that temperature, the volume will slowly contract toward B, or even past it, as the volume drifts toward the equilibrium state denoted by point C. This state is a hypothetical one for temperatures much below T_g as conventionally measured (i.e., cooling rates of a few tenths of a degree Kelvin per minute).

The fact that the glass is not at equilibrium raises some interesting questions in trying to describe its mechanical properties in terms of elasticity theory, because the latter assumes that a material is being deformed from a fixed reference state, i.e., that it is at equilibrium. In glasses, the reference state is steadily changing with time even in the absence of additional load or deformation.

Setting this point aside, the question is: how could one follow the change in volume from point A to point B and what is the effect of this densification on the physical properties?

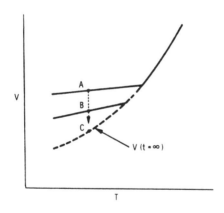

VOLUME-TEMPERATURE

Figure 2

From the WLF treatment of time scaling in polymer relaxation processes, it is known that anything which affects the fractional free volume f of a material leads automatically to a change in the time response through a change in the time-scaling parameter or time shift factor a, as shown by the second bullet in figure 3. Here B is the Doolittle constant and f_0 is f measured at a reference state. Thus if f changes with temperature as: $f_0 = f_0 + \alpha_f (T - T_0)$, where T_0 is a reference temperature, one obtains the WLF equation at the right.

The volume shrinkage from A-->B-->C is almost solely a reflection of the loss in f, which is related to the difference in the volume at some time V (e.g. point A of Fig. 2) and the volume at equilibrium or infinite time V_∞ (e.g. point C in Fig. 2). The rate of volume shrinkage has been extensively studied by Kovacs and by Kovacs et al. Denoting the fractional change in volume as δ, Kovacs showed that the rate of change of δ with ageing time was approximately first order, with a time constant τ; Kovacs et al. later showed that there is a spectrum of rates, so that $d\delta/dt$ is given by a summation over these rates. The problem then shifts to a need to calculate or measure δ.

- **RATE OF APPROACH TO EQUILIBRIUM**

$$\delta = \frac{V - V_\infty}{V_\infty}$$

$$\frac{d\delta}{dt} = \frac{\delta}{\tau_{eff}} \Rightarrow \sum^N \left.\frac{d\delta_i}{dt}\right|_T$$

HOW TO GET V_∞ IF T << Tg?

- **SHIFT FACTORS**

$$\ln a = B(f^{-1} - f_0^{-1}) \underset{\Delta T}{=} \frac{-(B/f_0)(T - T_0)}{(f_0/\alpha_f) + T - T_0}$$

KOVACS et al.

Figure 3

In order to calculate δ one needs first of all an analytic expression for the V-T response of the glass at the (hypothetical) equilibrium state (fig. 4). On the experimental side, one needs a direct measure of the volume state of the glass. This can be its time dependence during ageing/annealing as just described. Changes in volume can also arise with an applied strain and the magnitude of this change (and hence the magnitude of the change in physical properties) can also change with the strain field. That is, the value of δ for a given principal strain will depend on the degree of bi- and tri-axiality of the strain. Measurement of such volume changes accompanying the deformation should be made for a proper characterization of the response but unfortunately this is often done either crudely or not at all.

NEED:

1. V_∞-T IN GLASS STATE

2. DIRECT MEASURE OF VOLUME CHANGE WITH ANNEALING

3. DIRECT MEASURE OF VOLUME CHANGE WITH STRAIN AND STRAIN FIELD

Figure 4

The problem of defining an equilibrium volume temperature response and determining the resultant time scale parameter has been addressed recently by Curro, Lagasse and Simha. Simha and Somcynsky have developed an equation of state for liquids which are at equilibrium. It is based on a lattice model and relates reduced volume $\tilde{V} = V/V^*$, reduced temperature $\tilde{T} = T/T^*$ and reduced pressure $\tilde{P} = P/P^*$. The starred quantities are normalizing parameters which are material specific. At atmospheric pressure, a reduced equation of state results which relates V and T to the number of occupied lattice sites y. Empirically, it has been found that V^* and T^* can be determined by curve-fitting V-T data to the form $\ln \tilde{V} = A + B \tilde{T}^{3/2}$ and so the V-T curve in the liquid region can be described. This approach has been extended to liquids which are not at equilibrium, ie. glasses, by Curro, Lagasse and Simha, who pointed out that an isothermal change in volume with time of ageing t_a, i.e., $V(T, t_a)$, is equivalent to a change in y. Taking the fractional free volume to be equal to 1-y, they showed that it is possible to calculate V_{∞} vs T. Then, using directly measured V-t_a data, one can calculate δ, thence f, and therefore a_{t_a}, i.e., the last equation in figure 5. The term t^r_a is a reference ageing time.

$\tilde{V}, \tilde{T}, y$ RELATIONSHIP
$\ln \tilde{V} = A + B\tilde{T}^{3/2}$
 GIVES T^*, V^*, THEREFORE y
$\tilde{V}, \tilde{T}, y_{\infty}$ RELATIONSHIP

CLS:
$\tilde{V}(T, t_a) \rightarrow y(T, t_a)$
$f(T, t_a) \equiv 1 - y(T, t_a)$
$f_{\infty} = 1 - y_{\infty}$

$\text{LOG } a_T = \dfrac{B}{2.303}\left(\dfrac{1}{f(T)} - \dfrac{1}{f_{\infty}}\right)$, ABOVE T_g, GIVES B = 6.9

$\text{LOG } a_{T, t_a} = \dfrac{B}{2.303}\left(\dfrac{1}{f(T, t_a)} - \dfrac{1}{f(T, t_a^r)}\right)$

CURRO, LAGASSE, SIMHA

Figure 5

Using the JPL V-t_a data of Hong et al. (fig. 6), they were able to determine the value of y as a function of T and hence calculate both the equilibrium volume V_∞ and the occupied volume V_0 as a function of temperature for PMMA (fig. 7). From these data and the value of B for PMMA in the glass transition region as found by Schwarzl and Zahradnik, they were then able to calculate a_{ta}.

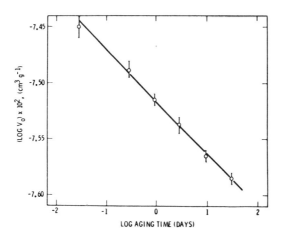

MEASURED VOLUME VS. AGING TIME AT ROOM TEMPERATURE, PMMA; CURRO, LAGASSE, SIMHA

Figure 6

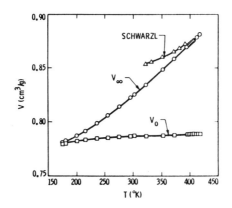

CALCULATED V,T RESPONSE; CURRO, LAGASSE, SIMHA

Figure 7

Figure 8 shows the excellent agreement obtained between the experimentally determined and calculated value of log a_{t_a}. Also shown (triangles) are the erroneous values of log a_{t_a} which are calculated if parameters appropriate to the liquid state are used in the WLF equation to calculate the shift factor.

Hence it now appears to be possible to obtain the (hypothetical) equilibrium V-T response of a glass and from this to calculate both the rate of physical ageing and the consequences of ageing insofar as they affect the time scale of the physical property responses. Obtaining the requisite data can be a tedious task and so we are trying to establish independent means of assessing the free volume content, in order to both test the theory and reduce the labor involved.

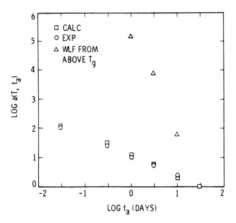

COMPARISON OF CALCULATED AND EXPERIMENTAL $LOG_{a_{t_a}}$ SHIFT FACTORS; CURRO, LAGASSE, SIMHA

Figure 8

Tsay et al. (1982) at JPL have found that by introducing spin-active molecules
into a polymer it appears to be possible to obtain a measure of the free volume.
Figure 9 shows the electron spin resonance (esr) signal from the probe molecule
TANOL (4-hydroxy-2, 2, 6, 6-tetramethyl-piperidine-1-oxyl) in PMMA. There is a
broad assymetric triplet, and on the high field side there can be a small narrow line
component. The figure shows the decay of this narrow line component with ageing time,
i.e., time after quenching the sample from the liquid state. This peak represents
a fast, easy motion; the broad peak represents a retarded motion. We can separate
out this narrow-line component by computer-assisted data reduction techniques and
show that it corresponds exactly with the probe response when the PMMA is in the
liquid state well above Tg.

Assuming that the existence and amplitude of the narrow line stem from the
presence of the free volume and are a measure of their magnitude, then the ratio of
the respective areas under these two peaks should be a direct measure of the free
volume. If this is so, then one can calculate log a_{ta} for the ageing PMMA. A
comparison of the fractional free volume measured by this technique and calculated
by Curro, Lagasse, and Simha in the manner just indicated shows that the agreement
is excellent (fig. 10).

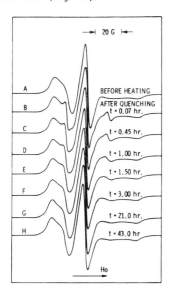

ESR DETECTION OF MOLECULAR RELAXATION IN QUENCHED
PMMA SAMPLE (PROBE B)
TSAY, HONG, MOACANIN, GUPTA (1982)

Figure 9

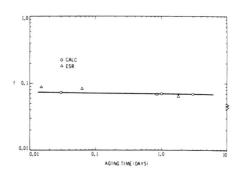

COMPARISON OF CALCULATED AND ESR-MEASURED FRACTIONAL FREE VOLUME IN PMMA
AT 296°K

Figure 10

We are assessing this new technique for measuring fractional free volume in several ways. For example, esr can be used to follow the extent of cure as measured by a microviscosity or highly localized viscosity. Here, as the cure develops, one observes that the fast component decreases with cure time. Thus as the polymer passes from the mobile liquid state to the glassy state, the fractional free volume decreases. We are also using esr to probe the molecular mobility within polysulfone (Tsay, Hong, Moacanin, and Gupta, 1981). On ageing, one finds changes in line shape similar to these already shown for PMMA (fig. 11). Moreover, careful analysis of the line shape shows that four separate relaxation times can be identified.

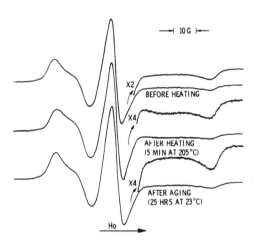

ESR DETECTION OF FREE VOLUME IN POLYSULFONE (TANOL PROBE)

Figure 11

We have been able to relate those relaxation times or their reciprocal (the frequencies of maximum absorption of energy) with the mechanical and dielectric relaxation peaks that one would conventionally measure using techniques such as the dynamic mechanical spectroscopy. We now plot each of these frequencies of the peak maximum against reciprocal temperature. Figure 12 shows the results for the dielectric data, other mechanical data, and the esr data. This plot is a typical way of comparing mechanical and other data. Clearly, the peaks we are picking up with the esr and their correlation times are exactly those which have previously been identified as the α, β, and γ and methyl rotation peaks in polysulfone.

Note that as the time scale of an experimental technique is changed, the peak maxima will occur at different temperatures. Thus an α peak measured at a time scale of, say, one second per cycle would represent what has been called the Tg of the material.

We feel that the gamma relaxation is related to the local reorientation of the phenyl ring, the β relaxation is the phenyl ring plus the sulfone and of course the α relaxation is always attributed to a large-scale segmental motion at Tg.

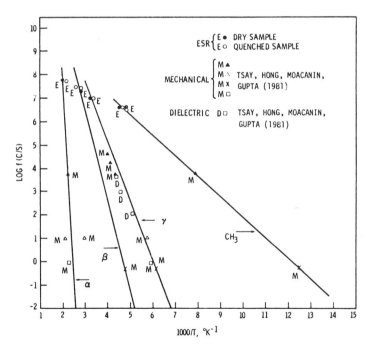

PLOT OF LOG F AGAINST 1/T FOR POLYSULFONE

Figure 12

We believe that it is the combined gamma and beta relaxations which are responsible, on a molecular scale, for the stress-strain response observed when a sample of polysulfone is deformed. These molecular rearrangements have very low activation energy, which would mean that they are able to respond very readily to deformations or loads on the sample. Figure 13 gives the transition temperature at the esr measured frequency for the four relaxation processes and their activation energies.

These measurements are being supplemented and expanded by Hong et al. at JPL, using birefringence measurements during the course of uniaxial tensile experiments. Such measurements can distinguish between a generalized lattice expansion, common to all solids, a local chain segment reorientation, and a full-scale chain reorganization and extension.

TRANSITION TEMPERATURES, ROTATIONAL FREQUENCIES, AND ACTIVATION ENERGIES FOR MOLECULAR RELAXATION PROCESSES IN POLYSULFONE

RELAXATION PROCESS	TRANSITION TEMPERATURE (°K)	ROTATIONAL FREQUENCY (MHz)	ACTIVATION ENERGY (KCAL/MOLE)
α	456	66	134
β	333	23	18.2
γ	294	12	12.1
CH_3	200	3.6	4.1

Figure 13

One particular point that I want to make in summary (fig. 14) is that, first, glasses are not at equilibrium. This has different kinds of consequences, one of which is the question of how one would describe the material. It means in a pragmatic sense that the volume changes which occur during physical ageing or as a result of straining will lead to changes in the nature of the material, and its reference state; these are changes which have implications for the validity of the equations of mechanics currently in use to analyze the response of glassy systems. Secondly, it is clear that we need to measure such volume changes explicitly. On the theoretical side we need an expression for the equilibrium PVT response. Fortunately there is at least one expression in the literature for this, which gives us hope. Thirdly, we are introducing what one might call a new technique to probe molecular motions during aging --- esr. It is new in the sense of its use for this purpose. Finally, in one case at least, it appears that the esr technique can be used to estimate fractional free volume and the changes of fractional free volume with ageing. This novel development must now be tested, as one of our future activities.

- **GLASSES NOT AT EQUILIBRIUM**

  CONSEQUENCES: PRAGMATIC

  THEORETICAL DESCRIPTION

- **NEED EQM. PVT INFORMATION/EXPRESSION**

- **INTRODUCING "NEW" TECHNIQUE TO PROBE MOTIONS AND AGING . . . ESR**

- **(IN ONE CASE) FRACTIONAL FREE VOLUME**

SUMMARY

Figure 14

REFERENCES

1. Cizmecioglu, M., Fedors, R. F., Hong, S. D., and Moacanin, J., The Effect of Physical Ageing on Stress Relaxation or Poly (Methyl Methacrylate), Polymer Eng. and Sci., 21, 942 (1981).

2. Kovacs, A. J., Fortschr. Hochpolym. Forschung, 3, 394 (1963) Adv. Polymer Sci., 3, 394 (1964).

3. Kovacs, A. J., Aklonis, J. J., Hutchinson, J. M., and Ramos, A. R., J. Polym. Sci. Polymer Phys. Ed., 17, 1097 (1979).

4. Curro, J. G., Lagasse, R. R., and Simha, R.; Use of a Theoretical Equation of State to Interpret Time-Dependent Free Volume in Polymer Glasses," J. Appl. Phys. 52, 5892 (1981).

5. Simha, R., and Somcynsky, T., Statistical Thermodynamics of Spherical and Chain Molecule Fluids, Macromolecules, 2, 342 (1969).

6. Hong, S. D., Chung, S. Y., and Fedors, R. F., Molecular Deformation and Stress-Strain Behavior of Diamino-diphenylsulfone-cured Tetraglycidyl-diamino-diphenyl Methane from Stress-Optical Studies. Org. Coatings and Appl. Poly. Sci. Proc., 48, 586-587 (1983).

7. Schwarzl, F. R., and Zahradnik, F., The Time-Temperature position of the Glass-Rubber Transition of Amorphous Polymers and the Free Volume, Rheol. Acta, 19, 137 (1980).

8. Tsay, F. D., Hong, S. D., Moacanin, J., and Gupta, A. Studies of Magnetic Resonance Phenomena in Polymers. I. The Effects of Free Volume and Segmental Mobility on the Motion of Nitroxide Spin Probes and Labels in Poly (methyl Methacrylate), J. Polymer Sci., Polymer Phys. Ed., 20, 763-772 (1982).

9. Tsay, F. D., Hong, S. D., Moacanin, J., and Gupta, A. Spin Trapping Studies of Tetraglycidyl-4,4'=diaminodiphenyl Methane (TGDDM) Cured with Diamino-diphenyl-sulfone (DDS), Polymer Preprints, 22, 231-232 (1981).

THE CHEMICAL NATURE OF THE FIBER/RESIN INTERFACE IN COMPOSITE MATERIALS

R. Judd Diefendorf

Rensselaer Polytechnic Institute
Troy, NY

INTRODUCTION

The first high modulus carbon fiber/epoxy resin composites were found to have low interlaminar shear strengths[1]. The question is where does the fracture occur, (Figure 1) and what factors affect the interlaminar strength. Carbon and aramide fibers are anisotropic and the properties transverse to the fiber axis are quite different than parallel to the axis[2]. In some cases such as Kevlar aramide or pitch precursor carbon fiber, the fiber itself (Zone A) may fail transverse to the fiber axis in an interlaminar shear test. Furthermore, the fibers are not homogeneous and have a skin (Zone C) which is quite different from the core (Zone A)[3-6]. The skin (Zone C) is usually quite graphitic and weak in shear, and failure can occur in this skin adjacent to the interface. Also, the skin can be hard to bond to, which leads to interfacial failure, (Zone D). In all cases, failure must also run through the continuous phase matrix. The matrix properties also are known to be different close to the interface (Zone E) as compared to the bulk (Zone B). Residual stresses arising from curing and cool down from processing temperature places the matrix in hoop tension around the fiber. These stresses are modified in an opposite manner by water absorption. Hence, the system is complex, and although fracture may occur near the interface, it does not need to occur exactly at the interface. This paper will emphasize the nature of the fiber structure, and the interaction that occurs at the interface between fiber and matrix.

POSSIBLE FRACTURE ZONES

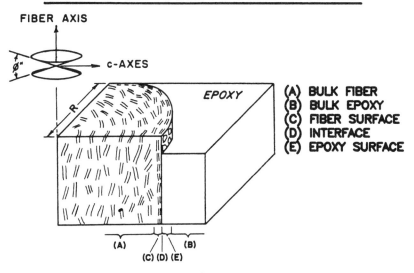

Figure 1

OPTIMUM INTERFACIAL BOND STRENGTH

When carbon fibers first came out, it was difficult to make a strong interfacial bond with the matrix. The problem was solved by treating the carbon fiber surface. However, the increased bond strength produced more brittle composites. Figure 2 shows that a bundle of fibers with no resin (left) will have about 70% to 80% of the average tensile strength determined on the single fibers. With a very strong interface (right) between fiber and resin in a composite the whole composite can fail when the very weakest fiber in the composite fails[7]. Tensile strength will be low, and failure catastrophic. A proper interfacial strength as compared to fiber strength will have multiple filament fractures, and the tensile strength will equal or often exceed the average fiber tensile strength. Depending on the application, the fiber/matrix interfacial strength should be adjusted. Hence for pressure vessels with low shear stresses, the interfacial bond strength could be quite low. For a structural application with significant shear stresses, the interfacial bond strength must be higher even though the resulting composite may be more brittle. While in the past the cost of qualification prevented application-optimized surface treatments of fibers, better composites could be made with better control of the interfacial properties.

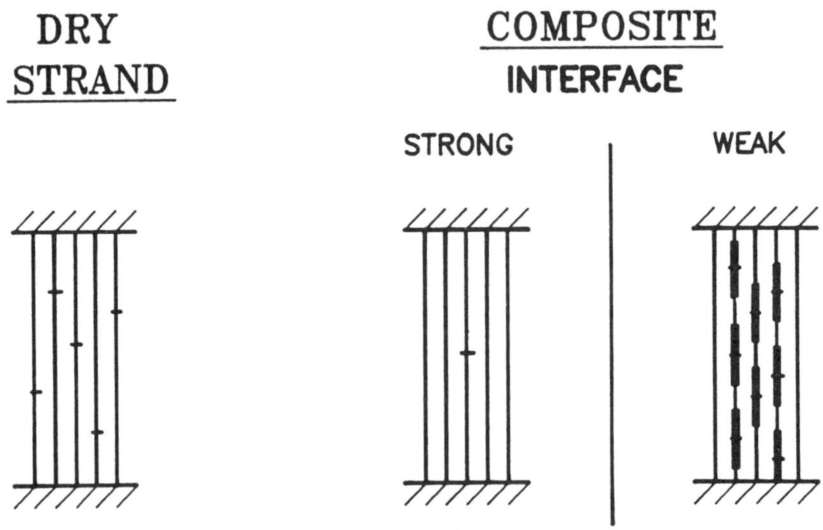

DRY STRAND

COMPOSITE INTERFACE

STRONG | **WEAK**

Figure 2

MICROSTRUCTURE OF CARBON FIBERS

The microstructure of carbon fibers (Figure 3a) is based on many people's work and especially some performed by my own students[3,4,5,6,8]. The axial structure resembles groups of undulating ribbons of graphitic based planes, and the modulus of the fiber depends on the amplitude-to-wavelength ratio of the ribbon. (The lower the amplitude and the longer the wavelength, the higher the modulus of the fiber is). For the cases of PAN and pitch precursor carbon fibers, higher modulus was obtained by heat treating to higher and higher temperatures. A problem is that the heat treatment also improves the orientation radially as well and cool down from high temperature causes residual stress cracks parallel to the fiber axis[8]. Also, when a high modulus fiber is made, the undulating ribbons become straighter and tend to form an onion-skin layer of good graphite at the fiber surface. This layer is hard to bond to, and has poor shear and transverse tensile strength. Figure 3b illustrates the undulating ribbon structure in a higher modulus carbon fiber[5]. The 1/2 wavelength is about 100Å. Close to the surface of this fiber, the structure would be almost perfectly oriented.

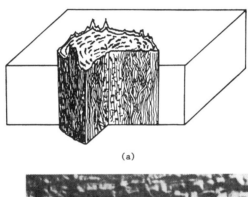

(a)

(b)

Figure 3

INTERLAMINAR SHEAR STRENGTH

The interlaminar shear strength of carbon fiber/epoxy composites decreases with increasing fiber modulus[1]. Even after surface treatment, the trend remains although the magnitudes are higher. The effect is not just a change in the mechanics of the problem caused by the increase in modulus, as the interlaminar shear strength for boron filament is high (Figure 4).

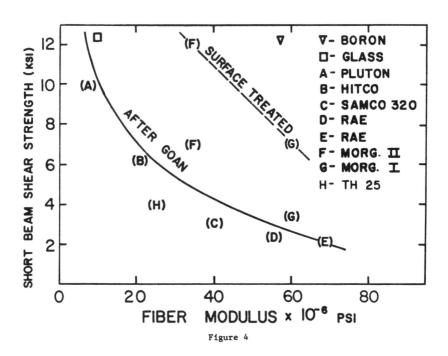

Figure 4

SURFACE ROUGHNESS EFFECTS ON INTERFACIAL BOND STRENGTH

The surface of carbon fibers does not significantly change with increasing fiber modulus at SEM magnifications (∼10,000X). Surface treatment appears to smooth the surface. However, TEM of replicas at higher magnifications (∼94,000X) shows quite different results[5]. Fiber surfaces get smoother with increasing modulus, and rougher with surface treatment. Figure 5a shows the pitted surface of a surface treated of a lower modulus fiber. By comparison, an unsurface-treated, very high modulus carbon fiber is quite smooth, Figure 5b. The shrinkage of the resin around the fiber upon processing would provide much better mechanical interlocking with the lower modulus/surface-treated fiber.

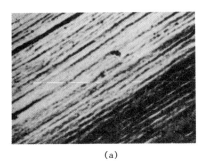

(a)

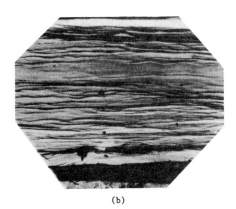

(b)

Figure 5

SURFACE ENERGY AND WETTING OF CARBON FIBERS

The difference in appearance of the surface of the carbon fibers with increasing modulus or after surface treatments indicates changes in the amount of graphitic basal plane and edge exposed on the surface. Figure 6 illustrates that the basal plane of graphite has a low surface energy and is hard to wet, while the edge has a high energy and is easy to wet (unless contaminated)[3]. The difficulty is that the surface is not monoenergetic, and the ease in wetting the surface will depend on the relative amount of edge and basal plane on the surface. Obviously, the higher modulus, unsurface-treated fiber would be harder to wet than the lower modulus surface-treated fiber.

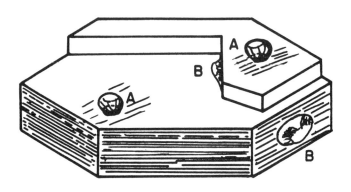

Wetting of Single Crystal Graphite

Figure 6

BASAL PLANE AND EDGE TOPOGRAPHY

The wetting of the surface depends on more than just the ratio of basal plane to edge. For example, for the same ratio of basal plane to edge, the wetting characteristics of a pitted surface will be different from a stepped surface, Figure 7. In fact, the wetting behavior for the stepped surface would be different if the liquid was advancing from left to right versus bottom to top. Part of this problem can be relieved by studying both the advancement and retraction of the liquid on the surface. However, wetting is only properly defined for a monoenergetic surface. For a stepped graphite surface, the surface may appear to be wetted because of the high surface energy steps, but the basal plane areas may not really be wetted. While wetting is not a sufficient condition for a good interfacial bond, it is a necessary condition.

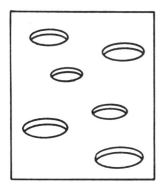
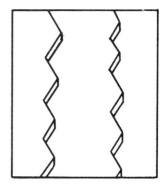

Figure 7

CRITICAL SURFACE ENERGY FOR WETTING

Zisman[9] has shown for surfaces, which have predominantly dispersive-type surface energies, that the critical surface energy for wetting can be obtained by extrapolating the measured contact angles for various dispersive liquids to perfect wetting. Hence, a liquid will completely wet the solid if it has a lower surface energy than the critical surface energy for wetting of the solid. For predominantly dispersive surfaces such as PTFE, the extrapolation works well with dispersive liquids, although more polar liquids give more scatter (Figure 8a)[10,11]. For a surface such as graphite with edges and basal planes, even more scatter should be expected. For highly perfect graphite surfaces prepared from highly annealed pyrolytic graphite, small variations in edge content can make significant differences, as can more polar liquids, Figure 8b[10,11]. Working with just the dispersive liquids, the surface energy for the graphite basal plane is quite low and it is hard to wet. (Independent calculations, based on elastic constants, give a value of 39 erg/cm^2)[12].

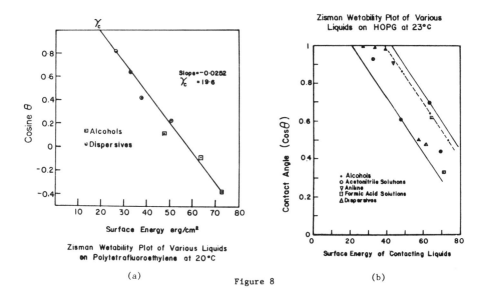

Figure 8

WETTING OF SURFACES BY POLAR LIQUIDS

Zisman[9] plots with slightly polar as well as dispersive liquids show more scatter. The polar liquids behave as if some of the surface energy doesn't count in terms of interacting with the surface to be wet. This could be explained by the polar part of the liquid molecule rotating away from the solid surface. By breaking the surface energy down into dispersive and polar components[13], we have found that only one-half the polar component of the surface energy is effective for predicting the critical surface energy for wetting[10,11]. Although this relation is not desirable from surface energy theories, the relation is so good (Figure 9) that there must be an underlying relationship.

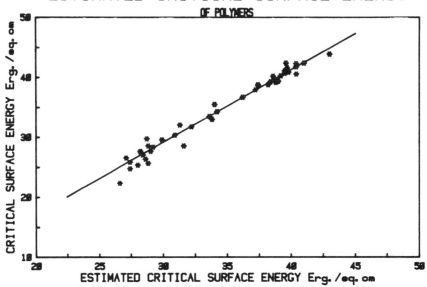

Figure 9

SURFACE ENERGY COMPONENTS

The idea of splitting the surface energy into dispersive and polar components has been effective in at least qualitatively explaining wetting behavior. This has been generalized to include a hydrogen bonding[14] and an acid/base interaction component[15]. (Logically, the hydrogen bonding term commonly used is improper.) However the results to be presented are based just on the dispersive and polar components as shown in the bottom equation of Figure 10. The approach was to describe the dispersive and polar components of a number of solid surfaces such as graphite with known liquids. Then these surfaces were used to determine the dispersive and polar components of resin systems such that the work of adhesion might be calculated.

THE GIRIFALCO–GOOD–FOWKES–KAEBLE–YOUNG EQUATION

GEOMETRIC MEAN RULE OF INTERFACIAL ENERGY

$$\gamma_{AB} = \gamma_A + \gamma_B - 2(\gamma_A \gamma_B)^{1/2}$$

ONLY DISPERSIVE AND POLAR INTERACTIONS

$$\gamma_{AB} = \gamma_A + \gamma_B - 2(\gamma_A^d \gamma_B^d)^{1/2} - 2(\gamma_A^p \gamma_B^p)^{1/2}$$

COMBINING THE YOUNG EQUATION

$$\cos\theta = -1 + \frac{2(\gamma_A^d \gamma_B^d)^{1/2}}{\gamma_L} + \frac{2(\gamma_A^p \gamma_B^p)^{1/2}}{\gamma_L} + \pi_e/\gamma_L$$

Figure 10

DETERMINATION OF DISPERSIVE AND POLAR SURFACE ENERGY COMPONENTS OF GRAPHITE AND PTFE

The contact angles for five liquids were measured on a freshly prepared highly oriented pyrolytic graphite (HOPG) surface and PTFE[10,11]. From the known surface energy components of the liquids, the surface energy components of the solids can be determined (Fig. 11). The slope of the plot is proportional to the polar component and the intercept to the dispersive component. The zero slope for PTFE shows it to have no polar component, and to have only dispersive surface energy. The values agree with the literature[16]. The values for HOPG are very similar to those obtained by Drzal[17] for HMU carbon fiber, which might be expected from microstructural results. In any case, the convenient flat HOPG surfaces appear to be useful for modeling carbon fiber surfaces.

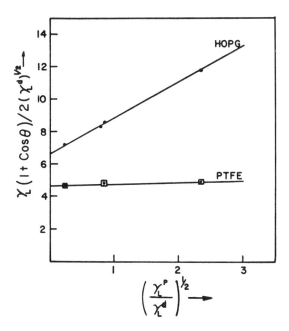

Determination of the Surface Energy Components of HOPG and PTFE at 20°C

Figure 11

CALIBRATED SOLID SURFACES FOR EPOXY RESIN STUDIES

The polar and dispersive surface energy components for a number of polymer and graphite surfaces were determined, Figure 12[10,11]. This provided a calibrated set of surfaces with quite different surface energy components for studying with resins.

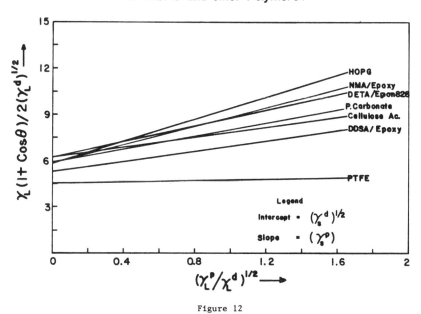

Figure 12

SURFACE ENERGY COMPONENTS OF EPOXY RESINS AND HARDNESS

The wetting of a number of epoxy resins and hardeners was studied on the calibrated surfaces. The corresponding surface energy components for resins and hardeners were calculated from these data. Some of the results are presented in Figure 13a[10,11]. Hopefully, the wetting and interaction of an epoxy/hardener mixture with a substrate could be calculated from the constituents. Unfortunately, the hardener, particularly amines, tends to interact or adsorb on the graphite surface, Figure 13b, and the wetting is mainly determined by the hardener[10,11]. While this makes the surface energy component approach of little value for direct calculation of wetting, it does point out when adsorption is occurring. It also provides information on how to maximize the surface interaction by acid/base phenomena.

SURFACE ENERGY COMPONENTS OF ANHYDRIDE CURED EPOXY RESINS

SUBSTRATE	Dispersive Component	Polar Component	Total Surface Energy
DDSA Cure:			
CY-179	29.2±0.6	2.9±0.7	32.1±1.4
MY-720	26.2±1.2	3.5±1.4	29.8±2.6
Epon-828	30.1±0.6	2.5±0.7	32.6±1.4
Epon-152	27.1±1.2	2.3±1.5	29.4±2.7
NMA. Cure:			
CY-179	34.3±1.2	10.2±1.4	44.4±2.6
MY-720	35.5±0.7	9.2±0.8	44.6±1.5
Epon-152	34.3±0.8	7.8±1.0	42.1±1.8

* All Values in Erg/sq.cm

Figure 13

SUMMARY OF COMPOSITE TOUGHNESS

composite toughness can be improved by increased axial tensile and compressive strengths in the fibers (Fig. 14). Secondly, the structure of carbon fibers indicates that the fiber itself can fail transversely, and different transverse microstructures could provide better transverse strengths. The higher surface roughness of lower modulus and surface-treated carbon fibers provides better mechanical interlocking between the fiber and matrix. The physical chemical nature of the fiber surface has been determined, and adsorption of species on this surface can be used to promote setting and adhesion. Finally, the magnitude of the interfacial bond strength should be controlled such that a range of composites can be made with properties varying from relatively brittle and high interlaminar shear strength to tougher but lower interlaminar shear strength. The selection would depend upon the application.

MATERIALS IMPROVEMENT

- **FIBER TENSILE AND COMPRESSIVE STRENGTH**
- **INTERFACIAL BOND**
    - FIBER SURFACE MORPHOLOGY AND CHEMISTRY
    - FIBER TRANSVERSE STRUCTURE
    - MATRIX CHEMISTRY
- **INTERPHASE**
    - DUCTILE OR HIGH ELONGATION
    - INTERMEDIATE MODULUS
- **MATRIX**

Figure 14

REFERENCES

1. Goan, J.C. and Prosen, S.P., "Interfacial Bonding in Graphite Fiber Composites Interfaces in Composites, ASTM-STP 4_2, ASTM 71st Session, San Francisco, CA, 1968, pp. 3-26.

2. Helmer, J.F., "Transverse Properties of Anisotropic Fibers and Their Composite: M.Sc. Thesis, Rensselaer Polytechnic Institute, Troy, NY, August 1983.

3. Butler, B.L., "The Effects of Carbon Fiber Microstructure on the Shear Strength of Carbon-Epoxy Composites", PhD Thesis, Rensselaer Polytechnic Institute, Troy, NY, September 1969.

4. Diefendorf, R.J. and Tokarsky, E.W., "High Performance Carbon Fibers", Polymer Engineering and Science, Vol. 15, No. 3, March 1975, pp. 150-159.

5. Diefendorf, R.J. and Tokarsky, E.W., "The Relationships of Structures to Properties in Graphite Fibers", AFML-TR-72-133, Parts I-IV, Oct. 1971 to Nov. 1

6. Johnson, D.J., Crawford, D., and Oates, C., "The Fine Structure of a Range of PAN-Based Carbon Fibers", Tenth Biennial Conf. on Carbon, Bethlehem, PA, June 1971.

7. Mullin, J., Berry, J.M., and Gatti, A., "Some Fundamental Fracture Mechanisms Applicable to Advance Filament Reinforced Composites", J. Composite Materials, Vol. 2, No. 1, June 1968, p. 82.

8. LeMaistre, C.W. and Diefendorf, R.J., "The Origin of Structure in Carbonized PAN Fibers", SAMPE Quarterly, Vol. 4, No. 4, 1973, pp. 1-6.

9. Zisman, W.A., "Constitutional Effects on Adhesion and Adhesions", in Symposium on "Adhesion and Cohesion", P. Weiss (Ed.), Elsevier, NY 1962.

10. Diefendorf, R.J. and Uzoh, C.E., "High Modulus Graphite Fiber Surface Modifications for Improved Intersection with Materials", 39th Semi-Annual Progress Report May 1980 - Sept. 1980, NASA/AFOSR Composite Structural Programs, Rensselaer Polytechnic Institute, Troy, NY.

11. Diefendorf, R.J. and Uzoh, C.E., "The Surface Energy of Anhydride-Cured Epoxy Resins", 43rd Semi-Annual Report, Dec., 1982, NASA/AFOSR Composite Structural Program, Rensselaer Polytechnic Institute, Troy, NY.

12. Riggs, D.M. and Diefendorf, R.J., "The Solubility of Aromatic Compounds", Proceedings of 14th Biennial Conference on Carbon, American Carbon Society, 1979, p. 407.

13. Kaeble, D.H., "Physical Chemistry of Adhesive", Wiley Interscience, NY, 1971, pp. 84-188.

14. Hansen, C.M. "Three Dimensional Solubility Parameter-Key to Paint Component Affinities: II and III, Dyes, Emulsifiers, Mutual Solubility and Compatibility and Pigments", J. Paint Tech., Vol. 39, No. 511, Aug. 1967, p. 505.

15. Fowkes, F.M. and Mostata, M.A., Ind. Eng. Chem. Prod., R&D, Vol. 17, No. 3, 1978

16. Wu, S., J. of Colloid and Interface Science, Vol. 17, No. 3, Oct. 1979, pp. 605-609.

17. Drzal, L.T., Mescher, J.A., Hall, D.A., "The Surface Composition and Energetics of Type HM Graphite Fibers", Carbon, Vol. 17, No. 5/A, 1979, pp. 375-382.

COMPOSITE PROPERTY DEPENDENCE ON THE FIBER, MATRIX, AND THE INTERPHASE

Lawrence T. Drzal

*Air Force Wright Aeronautical Laboratories
Nonmetallic Materials Division
Mechanics and Surface Interactions Branch
AFWAL/MLBM
Wright-Patterson Air Force Base, OH*

I. BACKGROUND

Most of the material mentioned in this conference represented commercially available prepreg and composites where the properties of the resin are fixed and the properties of the fiber are fixed, and, therefore, intrinsically the properties of the interface are fixed. In other words, they are in a condition, a necessary condition for good composite performance, where the adhesion between the fiber and matrix is at least an acceptable level if not an optimum level. The purpose of this workshop and the task of our group in the Air Force is substituting materials in composites - either matrices or fibers. Then we are really faced with the task of relooking at the problem of the interface between the fiber and matrix where the normal additive relationships between fiber and matrix, which in many aspects describe composite behavior from the mechanic's point of view, don't hold anymore. So our view, as prejudiced as it is, is that we really have to understand the behavior of the interphase as the way to combine matrix and fiber properties into composite properties.

I. BACKGROUND - INTERPHASE CONCEPTUAL MODEL

Figure 1 [1] illustrates what might be included in an interphase. From the mechanic's point of view we consider that the interphase between fiber and matrix is continuous in terms of transferring stresses between the fiber and matrix. If we look at that area under a high-resolution microscope then we could resolve these properties. We define an interphase of some finite distance which, depending on the material system, could extend from a few to a few thousand angstroms. In that interphase region starting from the matrix side, it can be seen that there can be physical differences. From the matrix side as we go to the interphase, we can have adsorbed or unreacted species congregated in the interphase region. We may have unwetted areas or voids or impurities congregating there. We may have surface chemical groups which would be entirely different from the bulk chemical material present at the surface of the fiber. We could have morphological or structural changes in the fiber at its surface. Finally we get to some point in the fiber where the local properties are equal to the bulk properties.

That is what I'm calling the interphase region. We are trying then to analyze this interphase in an environment which may be a thermal, a mechanical, or a chemical environment and which can come to the interphase through the matrix, along the interphase itself, or through the fiber. So you see it is a complex task, one that does not lend itself to easy analysis of single parametric models of adhesion and requires a phenomenological approach.

What I would like to do in this paper is to relate to you the results of various efforts conducted in our laboratories. I'm going to try to keep a degree of commonality throughout the presentations by talking about work that has been done on the same fiber, with the same matrix, and with one type of test.

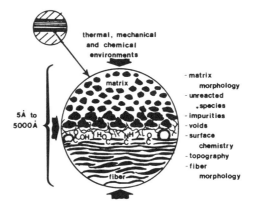

Figure 1.- Conceptual model of the fiber-matrix interphase in composite materials.

II. EXPERIMENTS

We have chosen, rather than to investigate a variety of different chemistries, to vary the matrix properties using one chemistry. The matrix chemistry we have chosen is an epoxy chemistry, in this case a difunctional epoxy, Epon 828, Shell Chemical Company, cured with meta-phenylene diamine. The cure cycle we have chosen is two hours at 75°C followed by two hours at 125°C. That gives us a matrix material that in the neat state has an initial modulus of around 500,000 psi with a strain to failure of around 6 to 7 percent and a fracture strength of around 12,000 to 13,000 psi. This matrix is intermediate in terms of properties between the brittle 5208 systems or 3501 systems and the rubber-toughened epoxy matrices.

Likewise for a fiber we have chosen to work with one fiber. It is the Hercules Type A fiber that has a 35 million modulus and a tensile strength of around 400,000 psi. We have obtained one large batch of the material, a portion of which was kept untreated. The rest of the material was surface treated with an oxidative surface treatment that was commercially available. Then we took a portion of the surface-treated material and had it finished, in other words we applied a thousand to two thousand angstroms of pure epoxy to the fiber surface in order to model an epoxy-compatible finished fiber. Those were the two components of our system.

Now depending on my outlook on a particular day, I could show you a composite fracture surface and I could pinpoint areas where you had interfacial fracture or had matrix fracture or had a combination of both. For an effort like this where an effort is being made to relate composite properties to the component material properties, composite testing is unsuitable. We have chosen to work with a single fiber test adapted from the metallurgists [2]. Advantage of the mismatch of strain properties between the fiber and the matrix is used to determine the interfacial shear strength.

A single graphite fiber is encapsulated in the matrix of interest, in this case our epoxy, in a tensile dogbone coupon specimen (Figure 2). That specimen is loaded in tension. Microscopically, in this sample at a low value of strain, the fiber will begin to break into fragments inside the matrix. As the strain on the sample is increased, the fiber in the sample will continue to break until you reach a point known as the critical transfer length where the shear stresses that build up on the fiber surface are no longer sufficient to break the fiber anymore. Extension of the matrix past this point causes no further breakage in the fiber. One can then by simple shear analysis relate critical transfer length, the fiber diameter and the fiber fracture strength at that length to an interfacial shear strength (Eqn. 1). In our experiments we usually use Weibull statistics. We just alter Eqn. 1 to reflect the Weibull statistics, and the calculation of the interfacial shear strength is made with the altered formula (Eqn. 2), which is our evaluation criteria for looking at interface properties.

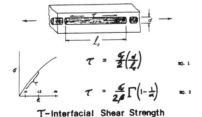

$$\tau = \frac{\sigma}{2}\left(\frac{d}{l_c}\right) \quad \text{Eq. 1}$$

$$\tau = \frac{\sigma}{2\rho}\Gamma\left(1-\frac{1}{\alpha}\right) \quad \text{Eq. 2}$$

τ-Interfacial Shear Strength

Figure 2.- The single-fiber interfacial shear strength test.

III. INTERPHASE: REINFORCEMENT SURFACE

The first part of our results will be looking at the fiber and the interface on the fiber side between fiber and matrix. For example if you analyze the fiber surface with some of the surface probes we have, the amount of oxygen on the surface undergoes about a twofold increase with surface treatment [3,4,5]. The surface oxygen seems to be of the carboxylic acid type. Likewise if you look at the total surface free energy, you see that it also goes up. The surface areas of the fibers do not change with surface treatment, at least not enough to reflect the doubling or tripling of interfacial properties that one would see with surface treatment. I would like to point out that when we expose the fiber to about 750°C in vacuum to a monolayer of hydrogen, we can remove all the oxygen down to about 3 percent. When we do that the polar component of the surface free energy and the total surface free energy of that fiber decrease without changing the fiber surface area.

The interfacial shear strength of those fiber specimens tabulated here shows that the untreated fiber gives a value around 4,000 psi. Surface treatment increases that to well over 10,000 psi. When we take those oxygen groups off the surface with the hydrogen treatment, the interfacial shear strength decreases. But it decreases to a level much higher than we get for the untreated fiber even though we remove oxygen groups to the point where we have 3 percent on the surface for this fiber and we have 9 percent for the untreated one.

That result puzzled us so we decided to look microscopically at the behavior of the fiber in the matrix. Figure 3 is just a collage of one particular fiber at its break in the matrix at about 400X with transmitted polarized light. After the fiber breaks, the matrix is placed under load and you see the stresses that develop with increasing strain. The fiber fragments tend to pull apart and immediately at very low levels of strain the entire fragment is in a state of stress. If you were to watch this in a dynamic sense, you could see the motion of alternating light and dark areas indicating that there is a nonsmooth progression of stresses as the fiber tends to move within the matrix. We have done some microtomy and ultrahigh resolution TEM work to look at the interface. We see for the untreated fiber that this material is entirely debonded from the matrix, and the fracture path goes between matrix and fiber interfacially but pulls apart fragments of the fiber surface from the untreated fiber.

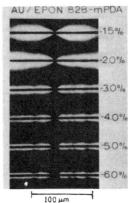

Figure 3.- Polarized transmitted light micrographs of an untreated (AU) fiber in the epoxy matrix with increasing strain.

Constituent Properties and Interrelationships 211

III. INTERPHASE: REINFORCED SURFACE - MICROGRAPHS OF THE SINGLE-FIBER TEST

Now contrast those results with the surface-treated fiber and you get quite different results (Figure 4). Here again the fiber is broken with increasing strain. A highly stressed area moved away from the break. This is the tip of an interfacial crack. Growing from the break in the fiber behind the crack tip is an area where there are high frictional forces responsible for the narrow band of intense photo-elastic behavior. If we take this surface-treated fiber and remove those oxygen groups, although the interfacial strength drops, we still get the same type of interfacial crack growth where we now get complete separation between fiber and matrix with no evidence of the fiber breaking up in its outer layers.

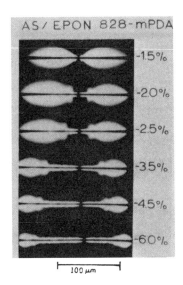

Figure 4.- Polarized transmitted light micrographs of a surface-treated (AS) fiber in the epoxy matrix with increasing strain.

III. INTERPHASE: REINFORCED SURFACE - SINGLE-FIBER INTERFACIAL SHEAR STRENGTHS

Interfacial shear strength is plotted as a function of surface oxygen in Figure 5. When we go from an untreated fiber to the surface-treated fiber, we get an improvement in interfacial shear strength. When we take the oxygen groups off the surface, we get a decrease, but the decrease still leaves us at a much higher value than we could get for the untreated fiber. Our explanation for this behavior is that two mechanisms are operating with these commercial oxidative surface treatments. The first part is the removal or etching away of the fiber surface, removing the weak boundary layer that is present on the fiber surface. The second step is the addition of the surface chemical groups. For this epoxy matrix we say that the surface treatment primarily creates a surface that can withstand higher shear loadings with the addition of surface chemical groups being a minor effect.

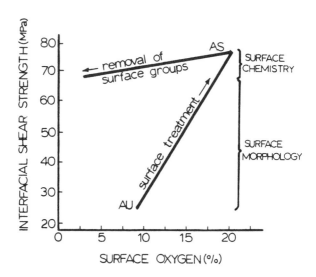

Figure 5.- Single-fiber interfacial shear strength plotted as a function of surface oxygen content.

IV. INTERPHASE: MATRIX

Let's look at the other side of the interface, and let's concentrate on the matrix side. The basic molecule we are working with is the DGEBA with N equal to about 0.1 which indicates we have some higher molecular weight species. The curing agent we are using is the meta-phenylene diamine. Its surface free energy is around 50 erg/cm^2 and the 828 surface free energy is around 42.3 erg/cm^2. Surface chemical effects for this particular system have been measured by using a pendant drop technique. By measuring the shape of the pendant drop of this material, one can determine if there is surface activity at the air-polymer surface. For this particular case as we increase the MPDA concentration, we have to go well beyond the stoichiometric point which is 14.5 phr before we start seeing any increase in surface free energy.

These epoxies are not homogeneous isotropic systems. They are quite inhomogeneous on a molecular scale. Figure 6 is a series of micrographs of epoxy samples at different amine-to-epoxy ratios [6,7]. The 14.5 phr is the ideal amount required to get complete reaction, 20 phr would be excess amine, and 10 phr would be a deficient amount of amine. The lower sample has been fractured in liquid nitrogen and then etched in a cold plasma. There are indications of heterogeneity indicated by the change in size and spacing with changing amine-to-epoxy ratio. The top specimens are microtomed samples which have been sliced into 600 angstrom thick sections. The shape of these osmium-stained dark areas and their concentration changes with the amine-to-epoxy ratio. This spherical morphology changes with composition.

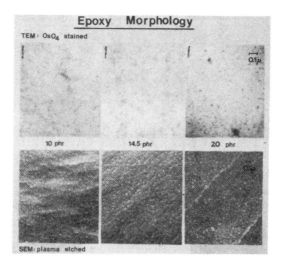

Figure 6.- Epoxy morphological changes occurring with alterations in amine-to-epoxy ratio. Top micrographs are ultramicrotomed sections which have been stained in osmium tetroxide. Bottom micrographs are epoxy samples fractured in liquid nitrogen.

IV. INTERPHASE: MATRIX - INTERFACIAL SHEAR STRENGTHS

We can alter the matrix side of the interface through the use of a finish. Graphite fibers are coated with about a thousand angstrom epoxy coating that is without the curing agent. When we compare just the surface-treated fiber with the surface-treated and finished fiber which only has a thousand angstroms of the pure epoxy resin, an improvement in interfacial shear strength is detected (Figure 7) [8].

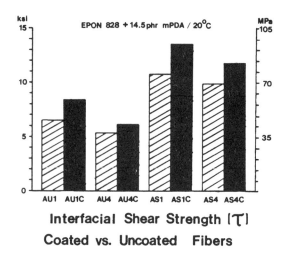

Figure 7.- Interfacial shear strength for carbon fibers with (C) and without epoxy finish.

IV. INTERPHASE: MATRIX - MICROGRAPHS OF THE CRITICAL-FIBER LENGTH TEST

Micrographs (Figure 8) of the fracture process that occurs show a third mode of failure. Whereas before we had an interfacial crack or actually frictional bonding now, with increasing strain after the fiber is broken, a matrix crack grows into the resin, not an interfacial crack.

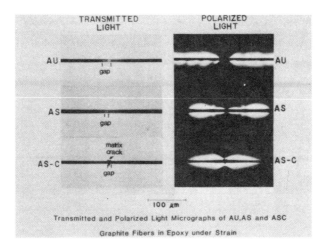

Figure 8.- Transmitted and polarized light micrographs of untreated (AU), surface-treated (AS), and surface-treated and finished (ASC) fibers under strain in an epoxy matrix.

IV. INTERPHASE: MATRIX - SCHEMATIC MODEL OF THE INTERPHASE

The explanation of what is happening here is the following (Figure 9). We are starting off with a fiber with a thin layer of finish at about 0 parts per hundred of amine. That fiber in turn is immersed in an epoxy matrix with 14.5 parts per hundred amine. In our cure cycle it takes about 50 minutes for gelation to occur. During that period amine from the bulk matrix is diffusing into the finish layer and an interphase region of some thickness is formed. The properties of this interphase region are now changing from the fiber surface on out into the bulk. As the amine diffuses it also reacts so a gradient of amine is formed going from the bulk to the surface of the fiber.

Some studies on model compounds where the amine-to-epoxy ratio has been varied show that when the amine content is reduced, the epoxy material tends to go up in modulus, at the same time decreasing in both fracture strength and fracture strain. In this case we are creating a brittle interphase between fiber and matrix which promotes better stress transfer because it has a higher modulus but also has a lower toughness. This interphase therefore promotes matrix fracture as opposed to interfacial fracture.

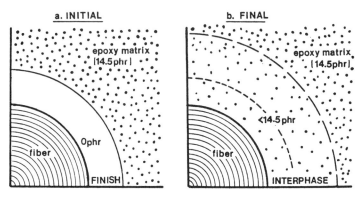

Model of Interphase Development During Cure Cycle

Figure 9.- Schematic model of a single finished graphite fiber in quarter view showing the initial and final states of the interphase.

V. INTERPHASE: HYGROTHERMAL EFFECTS

We have been looking at the surface treated and finished fibers at 20°C. Figure 10 shows what happens with hygrothermal exposure [9]. The Tg of the matrix we are using is about 166°C. The Tg of the coating, if we assume it has less than the stoichiometric amount of amine, about 7.5 parts per hundred, is about 60°C to 70°C.

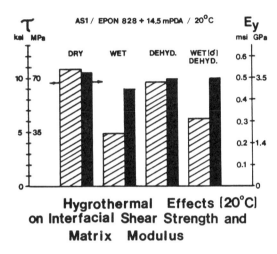

Hygrothermal Effects (20°C) on Interfacial Shear Strength and Matrix Modulus

Figure 10.- Interfacial shear strength and matrix modulus for a surface-treated fiber after hygrothermal exposure at 20°C.

V. INTERPHASE: HYGROTHERMAL EFFECTS - 125°C EXPOSURE

If we do our hygrothermal exposure at elevated temperatures (Figure 11) at least above the transition temperature of the coating, the coating acts like a rubbery material and acts to protect the interface. In the dry condition where we expose the surface-treated but uncoated fiber to moisture and then bring it down to room temperature and test it, interfacial shear strength has been reduced. Interfacial shear strength after drying is recovered. Now contrast that to the same fiber with the same matrix in the presence of that 1,000-angstrom finish layer of pure epoxy. A high value of dry properties is available initially, but in the wet state some loss is experienced nearly to the level of the uncoated fiber. After drying, no recovery is observed, but the ultimate or residual level is much higher than the level that is obtainable for the untreated fiber.

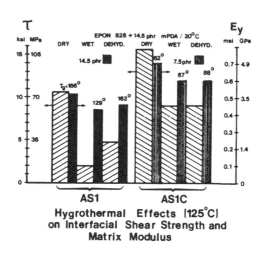

Figure 11.- Comparison of finished (AS1C) versus unfinished (AS1) fibers for interfacial shear strength and matrix and finish modulus after 125°C hygrothermal exposure.

V. INTERPHASE: HYGROTHERMAL EFFECTS - FRACTURED SURFACES PARALLEL TO FIBER AXIS

Finally let's look at how these interactions at the interface can affect the composite properties and how well the testing with a single filament test models composite properties. If you look at a composite in cross section, you see that the area where fibers come in contact is really a small proportion of the fiber diameter. Two thousand angstroms can be a significant portion of the interfiber distance in an actual composite. Composites were made from these three fibers: the untreated AU, the surface-treated AS, and the surface treated and coated ASC and at typical volume fractions with the same Epon 828-MPDA matrix system. Two different samples, both unidirectional about a tenth of an inch thick, were made. Compact tension specimens were fabricated with starter cracks in one case perpendicular to the fiber axis, in the other case parallel to the fiber axis [10]. The results are shown in Figure 12 for the untreated, surface treated, and surface treated and coated fibers. For the untreated fiber which had the lowest value of interfacial shear strength, the micrograph shows a lot of interfacial separation, clean fibers, and matrix without fibers in it. The intermediate surface treatment shows both situations, matrix failure as well as bridging in some areas of interfacial failure. Finally for the surface-treated and coated fiber, which had the highest degree of interfacial adhesion, there are no bare fibers. Every fiber that is there is coated with some degree of matrix.

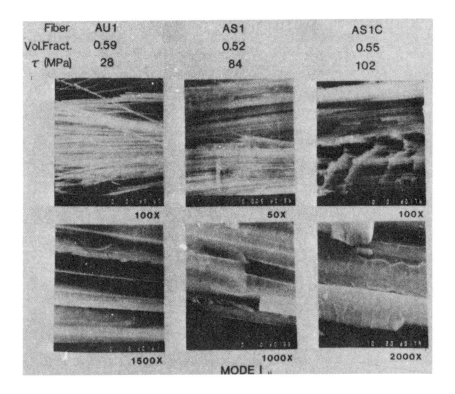

Figure 12.- Fracture surfaces of unidirectional composites with varying levels of adhesion parallel to the fiber axis.

V. INTERPHASE: HYGROTHERMAL EFFECTS - FRACTURED SURFACES PERPENDICULAR TO FIBER AXIS

Contrast the preceding results with fracture perpendicular to the fibers (Figure 13). The untreated fiber shows a very large degree of pull-out, where the pull-out lengths are on the order of half of what we measure in our single filament test. With surface treatment the pull-out length decreases. A very high degree of bonding gives almost planar fracture. The end of the fiber and the matrix surface are planar all along the entire fracture surface.

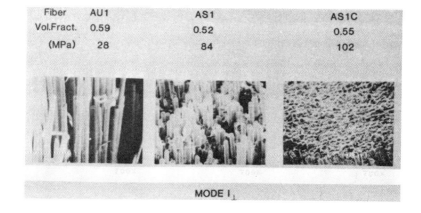

Figure 13.- Fracture surfaces of unidirectional composites with varying levels of adhesion perpendicular to the fiber axis.

VI. SUMMARY

In summary, I have tried to show that if one is going to consider altering the composite properties, for example, changing the toughness of the composite, one has to consider not only the matrix and the fiber but also the interface as well. What we would like to do ultimately is change the rule of mixtures from being purely additive between fiber and matrix properties to include the interface also as a design variable.

REFERENCES

[1] "Adhesion of Graphite Fibers to Epoxy Matrices. I. The Role of Fiber Surface Treatment," L. T. Drzal, M. Rich, and P. Lloyd, J. Adhesion, 16, 1-30 (1983).

[2] "A Single Filament Technique for Determining Interfacial Shear Strength and Failure Mode in Composite Materials," L. T. Drzal, et.al., Paper 20-C, 35th Annual Technical Conference, Reinforced Plastics/Composites Institute, SPI (1980).

[3] "The Surface Composition and Energetics of Type A Graphite Fibers," L. T. Drzal, Carbon, 15, 129-138 (1977).

[4] "The Surface Composition and Energetics of Type HM Graphite Fibers," L. T. Drzal, J. A. Mescher, and D. Hall, Carbon, 17, 375-382 (1979).

[5] "Graphite Fiber Surface Analysis through XPS and Polar/Dispersion Free Energy Analysis," L. T. Drzal and G. Hammer, Appl. Surf. Sci., 4, 340-355 (1980).

[6] "A Study of the Fracture Surface of Cured Epoxy Resin," L. T. Drzal, V. B. Gupta, and Y. L. Chen, Proceedings of the 41st Annual Meeting of the Electron Microscopy Society of America, 34-35, Phoenix, Arizona (1983).

[7] "A Modified Replication Technique to Study the Morphology of Cured Epoxy Resin," L. T. Drzal, V. B. Gupta, and R. Omlor, Proceedings of the 41st Annual Meeting of the Electron Microscopy Society of America, 36-37, Phoenix, Arizona (1983).

[8] "Adhesion of Graphite Fibers to Epoxy Matrices. II. The Effect of Fiber Finish," L. T. Drzal, M. Rich, M. Koenig, and P. Lloyd, J. Adhesion, 16, 133-152 (1983).

[9] "Moisture Induced Interfacial Effects on Graphite Fiber-Epoxy Interfacial Shear Strength," L. T. Drzal, M. Rich, and M. Koenig, Paper 4-G, 38th Annual Technical Conference, Reinforced Plastics/Composites Institute, SPI (1983).

[10] "Effect of Graphite Fiber-Epoxy Adhesion on Composite Fracture Behavior," L. T. Drzal, Proceedings of the 2nd U.S./Japan ASTM Conference on Composite Materials (1983).

NEWER CARBON FIBERS AND THEIR PROPERTIES

Roger Bacon

Union Carbide Corporation
Parma Technical Center
Parma, Ohio

INTRODUCTION

I will tell you about the newer carbon fibers that are either on the market or coming in the near future. Then I will discuss the structure of carbon fibers and describe how their properties depend on structure. Finally, I will describe how we get different types of structure by control of process parameters.

MATERIALS DEVELOPMENT GOALS

Some of the goals of composite materials development at Union Carbide are to achieve higher strength and stiffness, better damage tolerance, and better retention of properties at high temperatures. For some applications, dimensional stability of composites under rapid changes in temperature is required; here, modulus, thermal conductivity, and thermal expansion coefficient are important.

Better Composite Properties

- Strength
- Stiffness
- Damage Tolerance
- High Temperature Properties Retention
- Dimensional Stability (E, K_T, α_T)

KEY COMPOSITE PROPERTIES

Some key properties of composites are divided into fiber-dominated and resin-dominated properties. The fiber-dominated properties are mostly those measured in the 0 degree direction in a unidirectional composite. They include tensile strength, tensile modulus, compressive strength, thermal conductivity, and electrical conductivity. Shear strength is partly a fiber-dominated property (i.e., depends on the fiber surface) but it also is resin-dependent. Under resin-dominated properties, we include first the high temperature mechanical properties of the composite. Transverse strength, toughness, and (in part) longitudinal shear strength are resin-dominated properties.

Fiber-Dominated
   0° Tensile Strength
   0° Tensile Modulus
   0° Compressive Strength
   0° Thermal Conductivity
   0° Electrical Conductivity
   (Shear Strength)

Resin-Dominated
   High Temp. Mechanical Properties
   Transverse Strength
   Toughness
   (Shear Strength)

NEWER CARBON FIBERS

The current state-of-the-art carbon fibers are listed below. Carbon fibers tend to be classified by their modulus. The so-called high-strength type of fiber possesses a 33 million psi Young's modulus. The higher modulus fibers possess moduli of 55 million psi and 75 million psi.

Under the heading of "Newer Fibers", we see that much higher strengths are currently being realized in fibers of 35-to-37 million psi modulus. Next, there is an "intermediate modulus" fiber with 41 million psi modulus and excellent tensile strength. Finally, for applications requiring extremely high stiffness and excellent thermal stability, we have the ultra-high modulus fibers with stiffnesses of 100-to-120 million psi.

Fiber Type	State-Of-The-Art		Newer Fibers	
	Modulus	Strength	Modulus	Strength
High Strength	33 Msi	500 ksi	35 Msi	600 ksi
			37 Msi	750 ksi
Intermed. Modulus			41 Msi	700 ksi
High Modulus	55 Msi	350 ksi		
Very High Modulus	75 Msi	300 ksi		
Ultra High Modulus			100 Msi	325 ksi
			120 Msi	350 ksi

THE GRAPHITE LAYER PLANE

All carbon fibers are based on the graphite layer structure which is sometimes a very imperfect structure. The perfect graphite layer consists of a continuous "chicken-wire" network of carbon atoms. The layer planes are stacked together in a parallel arrangement. However, many carbon fibers, particularly the high strength varieties, have very small layer sizes and imperfect structures with missing carbon atom sites at which cross-linking can take place between the adjacent layers.

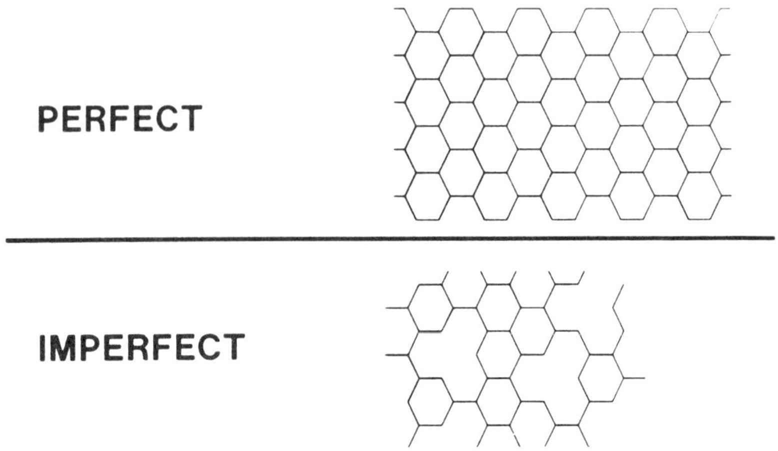

PERFECT

IMPERFECT

THE GRAPHITE CRYSTAL

The figure below represents the graphite crystal edge-on, showing a parallel stacking of the graphite layers. For the perfect graphite crystal, the Young's modulus is 148 million pounds per square inch in the layer plane direction. If the layers are imperfect and hence distorted and cross-linked, the modulus is lower. The shear modulus, on the other hand, is extremely low: only six-tenths of a million psi or even less. If it is an imperfect structure with cross-linking, one can expect a higher shear modulus. To estimate tensile strengths of graphite structures one must take into account these same differences in structure: very high tensile strengths are realized for pure tension parallel to the basal plane (as high as 3 million psi has been measured in graphite whiskers) but if appreciable shear stress is introduced, the strength is very much lower, being limited by the weak bonding between graphite layers.

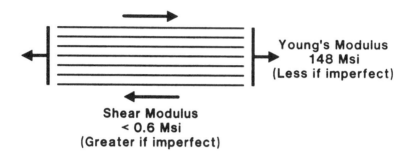

FIBER STRUCTURE

The main structural features of carbon fibers are indicated below. Orientation can be high or low depending upon whether the layers are straight and parallel to the fiber axis. Another important structural parameter is the crystallinity. If the linear dimensions of the perfect crystalline regions are large, the structure is said to possess a high degree of crystallinity; such regions tend to behave similarly to the perfect graphite crystal. Finally, defect content is very important, particularly with regard to fiber strength.

CARBON FIBER PROPERTIES ARE DETERMINED BY THEIR STRUCTURE:

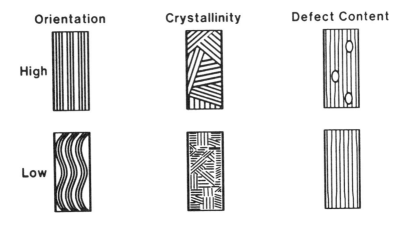

EFFECTS OF ORIENTATION ON PROPERTIES

During more than two decades, many workers have related carbon fiber structure to properties. I will give a qualitative listing of those properties which increase and those which decrease as one changes the level of each of the major structural parameters. We list below the effects of increasing preferred orientation. The longitudinal tensile strength, the longitudinal tensile modulus, the thermal conductivity, and the electrical conductivity all increase with increased orientation. The longitudinal negative CTE also increases; in other words, the coefficient of thermal expansion becomes more negative in more highly oriented (higher modulus) fibers. Properties which decrease with improved orientation are the transverse tensile strength and the transverse elastic moduli.

AS ORIENTATION IS IMPROVED, THESE CARBON FIBER PROPERTIES

INCREASE

Longit. TENSILE STRENGTH
Longit. TENSILE MODULUS
THERMAL Conductivity
ELECTRICAL Conductivity
Longit. Neg. CTE

DECREASE

Transv. STRENGTH
Transv. MODULI

EFFECTS OF CRYSTALLINITY ON PROPERTIES

Crystallinity affects many properties. Those which increase with increasing crystallinity include thermal conductivity, electrical conductivity and negative CTE measured in the fiber direction. (The thermal expansion coefficient in the transverse direction is always positive.) Oxidation resistance improves in a more crystalline fiber. Properties which decrease with higher crystallinity include the tensile and compressive strengths (assuming that the degree of preferred orientation is fixed.) The transverse strength and stiffnesses decrease, as does the longitudinal shear modulus.

AS CRYSTALLINITY IS IMPROVED, THESE CARBON FIBER PROPERTIES

INCREASE
THERMAL Conductivity
ELECTRICAL Conductivity
Longit. Neg. CTE
OXIDATION RESISTANCE

DECREASE
Longit. TENSILE STRENGTH
Longit. COMPRESSIVE STRENGTH
Transv. STRENGTH & MODULI
Longit. SHEAR MODULUS

EFFECTS OF FIBER DEFECTS

As you remove defects you improve just about everything: most notably, tensile strength. Thermal conductivity and electrical conductivity (both transport properties) are affected by certain kinds of defects. Oxidation resistance is affected by the presence of catalyst impurities.

AS DEFECTS ARE REMOVED, THESE CARBON FIBER PROPERTIES IMPROVE:

TENSILE STRENGTH
THERMAL Conductivity
ELECTRICAL Conductivity
OXIDATION RESISTANCE

CARBON FIBER MANUFACTURING PROCESS

The carbon fiber manufacturing process is outlined below. The precursor polymers may be cellulose (rayon), polyacrylonitrile (PAN), or pitch. Today, rayon is used only for very low modulus fibers. The precursor polymer is converted into a fiber by an extrusion process which may be either wet spinning, dry spinning, or melt spinning. The originally coarse fibers are stretched or drawn into finer fibers. The next step is to stabilize the fiber by oxidation in air at a temperature of approximately 400°C. The fiber is next carbonized to drive off most of the volatiles, leaving only the carbon behind. The carbonization temperature is between 1000°C and 2000°C. It is carried out in inert atmosphere. A final heat treatment, to achieve higher modulus or higher crystallinity, is performed at temperatures above 2000°C and sometimes as high as 3000°C.

Precursor Polymer	Cellulose (Rayon)
	Polyacrylonitrile
	Pitch (Mesophase)
Fiberize	Wet-, Dry-, Melt-Spin
	Draw
Stabilize	Oxidize
	Heat to 400°C
Carbonize	To 1500°C ($\pm$500°C)
	Inert Atmosphere
Heat Treat	To 2500°C ($\pm$500°C)

CONTROL OF STRUCTURE

We can control carbon fiber structure as follows:

1) Orientation is improved by fiber drawing or by restraining the fiber so that it can't shrink during the heat treatments. The precursor fiber structure helps to determine the degree of orientation in the final carbon fiber; starting with an oriented fiber structure one tends to improve orientation still further by heat treatment.

2) Crystallinity is largely determined ahead of time by the precursor chemistry. It is also strongly affected by the heat treatment, that is, by the final processing temperature.

3) Defect content is controlled by the purity of the raw materials and by the mechanics of fiber handling.

Structural Parameter:	Controlled By:
Orientation	1. Fiber Drawing
	2. Precursor Fiber Structure And Heat Treatment
Crystallinity	1. Precursor Chemistry
	2. Heat Treatment
Defect Content	1. Precursor Purity
	2. Process Handling, Etc.

FUTURE CARBON FIBERS

Sometime in the future I think we will see, for the high strength variety of carbon fiber, a one million psi tensile strength. For tough composites, carbon fiber producers are already providing strains-to-failure approaching 2 percent, and we expect to exceed this number in the future. In the case of high modulus fibers (55 million psi Young's modulus), 600 ksi tensile strength should be achieved. The ultra high modulus fibers will have a similar tensile strength but a lower strain-to-failure.

Fiber Type	Modulus	Strength	Applications
High Strength	40 Msi	1000 ksi	Structural
High Modulus	55 Msi	600 ksi	Structural
Ultra High Modulus	140 Msi	600 ksi	Space, Electrical

WORKING GROUP SUMMARY:
CONSTITUENT PROPERTIES AND INTERRELATIONSHIPS

R.F. Landel, Chairman

This session expressed considerable unease with the topic of toughness. The concept clearly has different meanings for different applications and different disciplines. A repeated call for guidance was made.

The framework for the session was set by noting the role of people working in this area relative to the hierarchy of activities from the end user back to the organic chemist. Thus, the structural engineering community defines the (generalized) loads which must be withstood; the composites designer/fabricator defines layups, the method of fabrication and the engineering material properties. The materials group, represented by those in Session II, then had the dual job of defining local response and failure (i.e., the micromechanics analysis of single and simple multiply systems) and translating the parameters or functions of micromechanics to molecular parameters of the matrix polymer chain (e.g., chemical units, chain topology, crosslink density) or to chain mobility, system morphology, or other structural features. The chemist then defines and creates appropriate molecular structures to accomplish the desired connectivity and morphology.

Of course this is a two-way street, at least up to the composites engineer. However, it is a vastly imperfect one. At each level there is a clear need for an input/output handoff of information. In particular, the resin chemist and materials property communities are clamoring for more explicit guidance from the next level up.

Session II therefore adopted a conceptual framework of considering the extent to which (or ease by which) one might be able to interpret or predict response. That is, assume one has a stated problem area and ask: What would you expect? How could it be analyzed? Do we have the appropriate tools and data? What characterizing parameters or functions appear, or have been omitted.

As an illustration, one currently assesses dry ply properties from, say, the fiber longitudinal and transverse moduli and strength, the matrix tensile and shear moduli and strength, and Poisson's ratio. These can be taken as input data to calculate stiffness but the prediction of strength can be far off the mark, and especially so with more ductile systems. Why? What is missing such that even a qualitative estimate cannot be made?

The resulting discussion led to the development of statements of problems which prevent us from carrying out such an analysis and translation, and also to comments that much relevant information was already available in each case but was not being used effectively. Clearly the widely varying disciplines involved in developing composites are not yet as successful as desired in communicating their needs and knowledge to each other.

Table I summarizes the requirements identified to address these problems. These are therefore recommendations for action. A discussion of each follows. Note that fibers were not discussed per se, and that processing was alluded to several times but not addressed explicitly in this session.

TABLE I

CONSTITUENT PROPERTIES AND RELATIONSHIPS:
REQUIREMENTS AND RECOMMENDATIONS

1. Need a program goals/requirements statement

2. Need a better definition of toughness: application, test mode, test commonality

3. Assess limitations of current micromechanics models/procedures

4. Assess currently known resin toughening mechanisms and determine their relative importance to matrix resins

5. Need methods to define, in situ, the properites of the interphase region

6. Assess thermodynamic compatibility of any new resin with the fiber

7. Study model polymer matrices, including ideal as well as simplified/practical systems

1. **Goals/requirements statement**: Concern was expreseed over the need for a clearer statement of the goals of the program and the real requirements of the airframe. More directly pertinent to the session was a perceived lack of direction or lack of targets to aim for from the fracture community.

2. **Better operational definition of toughness**: This requirement was in part a more explicit statement of previous requirement. The quantitative meaning of toughness and damage tolerance seems to shift from application to application. Considerable and repeated concern was expressed over the apparently evanescent nature of these targets. The opinion was strongly voiced that the search for a universally applicable answer should be replaced with a series of more narrowly defined targets and that Session I should translate such targets into clearer statements of requirements for Session II. For example, impact resistance on thick and thin sections and the potentials for crack propagation represent four different situations. Separate solutions to the problems may be easier to attain.

In addition to the need for operational definitions, there is a need to assess which of the various fracture toughness tests ought to be employed for resins and composites and then to adopt appropriate ones.

When reporting program results, contractors should use some consistent set of tests to define the properties or changes in properties with formulation, such that there would be at least some common means of comparing results.

3. **Assess limitations of current micromechanics models/procedures**: Most numerical models take the matrix material response as linear elastic. As such, they can predict small deformation response (e.g. modulus and expansion coefficients), but they cannot predict failure. What is needed is to incorporate the ability to handle viscoelasticity, nonlinear multiaxial response, and better failure criteria. Various ways of accomplishing the first two have been developed, and failure criteria represent in principle merely some predefined cap on the response. Two elements are missing, however.

First, there is a need to employ existing procedures (codes) to analyze various test geometries or damage situations and do a parametric study or sensitivity analysis to assess the importance of various assumed but realistic material behaviors. In this connection, the reminder was made that crack tip processes occur on very short time scales, on the order of microseconds, and this should be factored into input data or interpretation.

Second, there is, or appears to be, an almost total dearth of experimental data on nonlinear multiaxial behavior of matrix materials and failure behavior under multiaxial loads. Hence, analyses of their response at conditions approaching rupture or crack initiation cannot be carried out at this time, even where the numerical techniques are said to exist.

4. **Explicitly assess currently known resin toughening mechanisms and determine their relative importance to matrix resins**: At least five means of toughening resins and several for composites are known. The relative importance of these to a composite matrix is not known. Thus, certain amorphous thermoplastics craze or can be induced to craze (e.g., with the addition of rubber particles). But it is not clear that thermosets craze, or, should they be shown to craze, that the extent of crazing would be significant enough to impact the resistance to crack propagation. Similarly, rubber toughening is accompanied by a large deformation zone ahead of a crack tip, but the constraints of the fiber will restrict the size of the disturbed zone and reduce the in situ effectiveness. Correspondingly, crystalline polymers can be tough, but will the presence of the fiber or the bonding layer induce an unfavorable size or orientation of the crystalline region, destroying the toughness? In short, studies are needed on the extent to which toughening mechanisms useful in bulk resins will carry over to the in situ matrix resin.

One suggested characterization technique was that of determining the biaxial stress-strain response up to failure.

5. **Method to define the interphase region in situ**: There is no accepted technique for characterizing the polymer in the interphase region for graphite fiber systems. Without this the physical properties of the region cannot be interpreted in terms of molecular origin, even if they could be measured, and with neither available there are no real or assumed-but-reasonable input data for use in micromechanical models.

6. **Assess thermodynamic compatibility of any new resin with the fiber**: When new resin systems are being investigated their thermodynamic ability to interact effectively with the fiber surface or surface coating should be determined, lest promising candidates be unnecessarily discarded.

7. **Model polymer systems – idealized, simplified real**: Just as micromechanics models need careful assessment and employment, it was felt that the study of model matrix materials such as polymethyl methacrylate or polyethylene terephthalate would fulfill a real need in helping to distinguish the roles of molecularly related parameters such as backbone moieties or network topology. The time dependence, nonlinearity, failure behavior (e.g., brittle-ductile transition), and other such properties are well characterized and the molecular origins of the responses are reasonably well understood. Thus, there exists a precalibrated tool, as it were, to examine the effects of such parameters on composite behavior rather more closely than has been possible to date. Similarly, with such well-characterized polymers, one has hard input data to use in exercising the micromechanics models and checking the extent to which they can predict the response of real materials.

Part III
Matrix Synthesis and Characterization

The information in Part III is from *Tough Composite Materials,* compiled by Louis F. Vosteen, Norman J. Johnston, and Louis A. Teichman, NASA Langley Research Center, Hampton, VA, 1984. The report is the proceedings of a workshop sponsored by NASA Langley Research Center, May 1983.

DEVELOPMENT OF A HETEROGENEOUS LAMINATING RESIN

Rex Gosnell

*Narmco Materials, Inc.
Anaheim, California*

OBJECTIVE

The early part of this program was directed toward the feasibility of toughening the common types of matrix resins such as Narmco 5208 by utilizing a heterogeneous additive. The more significant latter part of the program evolved into a study of some basic concepts and principles in the toughening of matrix resins for advanced composites. An effort was made to determine the "why" and "how" of some approaches to toughening and to make an assessment of the sacrifices that must be made in the overall mechanical properties of a matrix resin.

POLYBLEND EPOXY ADHESIVES

At the outset of this program, it was thought that the technology involving heterogeneous resins that have been successful with adhesives might be applicable to resin matrices. A number of commercial aircraft adhesives utilize a blend of a discontinuous dispersion of elastomer in a continuous phase of a cross-linked resin system. These polyblends were quite successful in some early adhesive systems such as nitrile-phenolics. The improvement in impact capability of such resins led to the thought that this might be a viable approach for composite matrix resins.

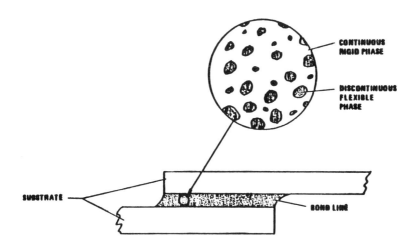

MECHANICAL BEHAVIOR OF POLYMERIC MATERIALS

Polymeric materials have a wide range of stress-strain behavior. Examples are shown. Of these, the last type seems to be the apparent need for improved tough resin matrix resins. As stated in the objective, the question is how you can improve this relationship of stress and strain in such a way that the overall performance of the composite will be attractive for advanced composites.

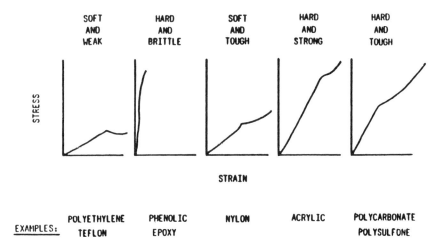

ELEMENTS OF MELT TRANSFER PROCESS FOR HIGH-MODULUS FIBER

Ideally, any modifications of resin systems would be compatible with existing prepreg processes, although this was not a primary consideration in this program. For example, some combinations of materials were very difficult to handle even in the laboratory, and would be totally unsuitable for existing prepregging methods. Information gained from these difficult systems might shed some light on other more viable approaches.

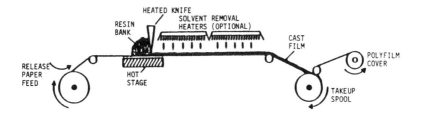

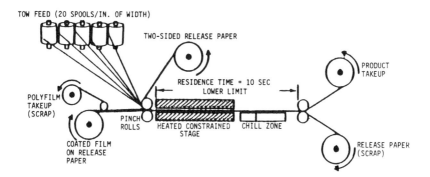

Tough Composite Materials

NARMCO COMPOSITE IMPACT SCREENING TEST

Narmco presently uses an impact screening test for assessment of the effectiveness of toughening of resin systems. This test was used in this program to determine relative toughening capability of the various approaches. The test is essentially a falling dart; the data treatment is discussed in the figure. A smaller number means less impact damage.

- 3K-70P/C-3000 FABRIC SOLUTION COATED
- 6" x 6" x 6 PLY LAMINATE
- DAMAGED AREA MEASURED AT 10, 20, 30 AND 40 IN. LB.
- GARDNER DAMAGED AREA FUNCTION QUANTITATIVELY DESCRIBES DAMAGE SUFFERED BY TEST LAMINATE AFTER IMPACT (REPRESENTING LOADINGS UP TO 900 IN. LB./IN. FOR NOMINAL .045" THICK LAMINATES)
  - REPORTED NUMBER IS WEIGHTED SUMMATION DERIVED BY INTEGRATING THE REGRESSION CURVE OVER THE SPECIFIED IMPACT ENERGY LEVELS
  - COMPUTED BY IN-HOUSE DEVELOPED EQUATION WHICH NORMALIZES THICKNESS AND REDUCES DATA SCATTER BY LEAST SQUARES LINE FIT

IMPACT DAMAGE VALUE

The value of this impact damage is correlated to ETI test values.

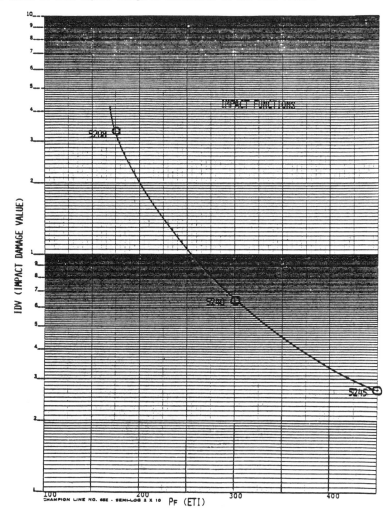

SHORT BEAM SHEAR

Retention of basic mechanical properties was monitored by using a standard short beam shear test. Early in the work, it was learned that this test gave reduced values when most of the toughening approaches were incorporated in the resin matrix resins.

12 PLY 4" x 6" LAMINATE

3K-70P/C-3000 FABRIC

SPAN/DEPTH = 4:1

WET TESTS AFTER 40 HOUR WATER BOIL

RT 200°F 270°F WET & DRY

PLASTIC FAILURES NOTED

"ALL-EPOXY" SYSTEMS

Using a simple epoxy formulation, several potential toughening agents were evaluated by the impact resistance and the short beam shear at RT, 200°F and 270°F, both dry and wet. Comparative data is also shown for Narmco 5208 and Narmco 5245C.

Hydroxyl-terminated polyethers, hydroxyl-terminated polyesters, ABS, and two CTBNs all produced improvement in impact value, but at a considerable sacrifice in 270°F wet short beam shear. Of all the material tested, the polycarbonate was the most effective in improving impact with the least sacrifice at the 270°F wet value. Note that Resicure #4 also improved impact strength with a similar reduction in 270°F wet short beam shear strength.

The polycarbonate was studied further in the program. It showed the most attractive improvement in impact value. (Smaller numbers are better.)

(2 Hrs. @350°F Cure)

Ingredients	1	2	3	4	5	6	7	8	8A	5208	5245C
Epoxies											
Den 439	11	11	11	11	11	11	11	11	11	-	-
FCI 98-180	4	4	4	4	4	4	4	4	4	-	-
Ciba MY-720	50	50	50	50	50	50	50	50	50	-	-
Hardener											
D.D.S.	21	21	21	21	21	21	21	21	21	-	-
Catalyst											
Resicure #4	.3	.3	.3	.3	.3	.3	.3	.3	-	-	-
Filler											
Cabosil	4.5	4.5	4.5	4.5	4.5	4.5	4.5	4.5	4.5	-	-
Toughening Agent											
Hydroxyl Containing Polyether (High M.W.)	5	-	5	-	-	-	-	-	-	-	-
Hydroxyl Terminated Polyester (Low M.W.)	8	10	-	-	-	-	-	-	-	-	-
Polycarbonate	-	-	-	4	-	-	-	-	-	-	-
ABS (K-2945)	-	-	-	-	6	-	-	-	-	-	-
CTBN (I)*	-	-	-	-	-	6	-	-	-	-	-
CTBN (II)*	-	-	-	-	-	-	6	-	-	-	-
Laminate Mechanical (3K70P)											
Short Beam Shear (KSI):											
R.T. (Dry)	13.3	10.6	12.6	12.1	12.6	12.0	12.2	12.0	10.2	13.7	12.5
200°F (Dry)	10.2	7.7	9.6	10.2	9.3	9.5	9.1	9.9	9.8	10.7	11.2
270°F (Dry)	6.8	6.2	6.9	8.5	7.3	7.7	7.7	7.0	7.5	9.6	9.4
RT (Wet)**	10.0	9.8	11.3	10.0	10.6	10.0	10.1	10.6	10.3	12.1	12.0
200°F (Wet)	6.6	5.5	6.7	7.7	6.4	6.8	6.4	7.5	8.4	9.0	10.1
270°F (Wet)	2.7P	3.6P	3.7P	4.7P	3.8P	3.9P	3.6P	4.6P	6.2	7.5	8.6
Impact Resistance	415	600	1300	303	586	412	782	915	1964	3300	270

*CTBN (I) = CTBN 1300 prereacted with an epoxy novalac.
CTBN (II) = CTBN 1300 prereacted with Ciba MY-720.

**Wet = 40 hour water boil.

MORE "ALL-EPOXY" SYSTEMS

The effect of amine hardener on mechanical properties and impact resistance was studied, concentrating on the polycarbonate as the toughening agent. In example number 3, a very good impact number was observed, but a severe penalty was imposed on the mechanical properties. Anchor 1482 is a eutectic mixture of aromatic amines which showed some promise, but again, the poor 270°F wet strength is a sign that a heavy sacrifice is taken in composite mechanical properties to achieve the good impact value.

(2 Hrs. @350°F Cure)

Ingredients	1	2	3	4	5	6	7	5208	5245
Epoxies									
MY 720	-	-	50.0	51.0	51.0	51.0	51.0	-	-
ERL 0510	5.0	5.0	5.0	5.0	5.0	5.0	5.0	-	-
DEN 439	5.0	5.0	8.7	7.0	7.0	7.0	7.0	-	-
RCI 98-180	-	-	-	3.0	3.0	3.0	3.0	-	-
XU-276	67.7	57.0	-	-	-	-	-	-	-
Hardeners									
LSU-931	-	22.5	24.0	23.0	23.0	23.0	23.0	-	-
XU-205	-	-	-	7.0	7.0	7.0	7.0	-	-
DDS	16.0	-	-	-	-	-	-	-	-
Anchor 1482	-	3.2	5.0	-	-	-	-	-	-
Catalyst									
Resicure #4	0.3	0.3	0.3	0.3	0.3	0.3	0.3	-	-
Filler									
Cabosil M-5	3.5	3.5	3.5	3.0	3.0	3.0	3.0	-	-
Toughening Agent									
Lexan	4.0	4.0	4.0	-	4.0	4.0	4.0	-	-
DER 669	2.0	3.0	3.0	-	-	3.0	3.0	-	-
Photomer 4127	-	-	-	-	-	-	1.0	-	-
Laminate Mechanical (3K70P)									
SBS									
RT Dry	10.5	11.0	9.6	11.0	10.2	12.1	11.8 (12.0)	13.7	12.5
200°F Dry	7.5	6.5	8.0	9.1	9.1	9.4	9.4 (9.8)	10.7	11.6
270°F Dry	2.8	1.4	6.4	6.4	6.2	6.7	7.1 (8.1)	9.6	9.4
RT Wet	-	-	10.0	10.0	10.4	10.6	10.6 (10.6)	12.1	12.0
200°F Wet	-	-	6.8	6.5	6.3	6.9	6.7 (7.6)	9.0	10.1
270°F Wet	-	-	3.9	3.0	2.8	2.9	2.5 (5.6)	7.5	8.6
Impact Resistance	784	1084	144	1560	655 (624)	379 (562)	632	3309	270

Note: Values in parentheses obtained on post-cured samples (4 Hrs. @ 400°F)

Matrix Synthesis and Characterization

BISMALEIMIDE/EPOXY SYSTEMS

The toughened bismaleimide/epoxy systems did not show the severe loss in hot-wet strength that was observed in the all-epoxy systems. Although room temperature values were somewhat lower than the epoxies, the temperature profile was more flat and the 270°F wet values were improved over straight epoxies. Impact values were respectable in the 300-700 range. (Lower values indicate less damage volume.)

(2 Hrs. @350°F Cure)

Ingredients	1	2	3	4	5	6	7	8	9	10	5208	5245C
Bismaleimide K-353	50	50	24	24	24	24	24	24	24	24	-	-
Epoxies												
Ciba MY 720	30	30	-	-	-	-	-	-	-	-	-	-
Ciba 0510	-	-	45	45	45	45	45	45	45	55	-	-
XU-276	-	-	-	-	-	-	-	-	-	14	-	-
Hardeners												
DDS	10	10	24	24	24	24	24	24	24	14	-	-
Catalysts												
Resicure #4	.3	.3	.3	.3	.3	.3	.3	.3	-	-	-	-
ETPI	-	-	-	-	-	-	-	-	.2	.3	-	-
Filler												
Cabosil	-	-	4.5	4.5	4.5	4.5	4.5	4.5	2	2	-	-
Toughening Agent												
Hydroxyl Containing Polyether (High M.W.)	8	5	-	-	-	-	-	-	-	-	-	-
Hydroxyl Terminated Polyester (Low M.W.)	2	-	-	-	-	-	8	-	-	-	-	-
ABS	-	-	-	6	-	-	-	4	6	6	-	-
Polycarbonate	-	5	-	-	-	-	-	2	-	-	-	-
CTBN/Epoxy I*	-	-	-	-	6	-	-	-	-	-	-	-
CTBN/Epoxy II**	-	-	-	-	-	6	-	-	-	-	-	-
Laminate Mechanical (3K70P)												
SBS (KSI)												
RT (Dry)	7.9	7.1	7.7	7.5	8.0	8.7	8.1	8.5	10.5	13.0	13.7	12.5
200°F (Dry)	7.2	6.1	8.1	8.2	8.2	8.7	7.9	8.3	9.5	11.5	10.7	11.2
270°F (Dry)	5.9	5.1	9.4	7.1	8.1	8.4	7.8	8.6	7.5	8.9	9.6	9.4
RT (Wet)***	6.7	8.9	8.5	9.0	8.7	8.3	8.6	9.4	9.5	13.6	12.1	12.0
200°F (Wet)	5.1	4.6	8.0	6.7	6.9	7.4	6.7	7.8	6.9	9.0	9.0	10.1
270°F (Wet)	3.7	2.9	6.9	4.6	5.5	5.2	4.5	5.3		5.7	7.5	8.6
Impact Resistance	261	305	738	706	832	679	622	641	678	790	3300	270

*CTBN 1300 prereacted with epoxy novalac.
**CTBN 1300 prereacted with MY-720
***40 Hr. water boil

CONCLUSIONS

1. The use of damage volume as a guide for measurement of impact resistance appears to be a valid determination.

2. Short beam shear is a good test to determine the effect of toughening agents on mechanical properties.

3. Rubber toughening results in improved laminate impact strength, but with substantial loss in high temperature dry and wet strength.

4. In the all-epoxy systems, the polycarbonate toughening agent seemed to be the most effective, although hot-wet strength is sacrificed. ABS was not as effective.

5. In general, the toughened all-epoxy systems showed better damage tolerance, but less hot-wet strength. Toughened bismaleimides had better hot-wet strength.

MODIFIED EPOXY COMPOSITES

W.J. Gilwee
NASA Ames Research Center
Moffett Field, CA

The potential use of graphite-fiber/resin matrix composites to achieve weight savings in aircraft and space applications is well documented (1). The design requirements for this type of composite include high strength, stiffness, impact resistance, and resistance to burning. The use of reactive liquid rubber to improve the toughness (impact resistance) of epoxy resin composites has been reported by a number of investigators (2-5). The two-phase system of a brittle epoxy resin phase and rubber phase is believed to increase the impact strength by means of a crack-terminating mechanism.

In this investigation, which is part of a screening program, we have studied the properties of a rubber-modified experimental epoxy resin as well as a standard epoxy as composite matrices. In addition, a brominated epoxy resin was used in varying quantities to improve the fire resistance of the composite.

OBJECTIVES
CARBON FIBER/RESIN COMPOSITES

- 350°F OR LOWER CURE TEMPERATURE
- HOT/WET PROPERTIES
- IMPROVED TOUGHNESS OVER CURRENT EPOXY RESIN
- LOW FLAME PROPAGATION
- LONG SHELF-LIFE PRE-PREG

The experimental resin was tris-(hydroxyphenyl)methane triglycidyl ether, known as tris epoxy novolac (TEN). The standard epoxy resin used was tetraglycidyl 4,4'-diaminodiphenyl methane (TGDDM). The brominated epoxy was bisphenol A polymer with 50% by weight Br.

The above resins were modified with carboxyl-terminated butadiene acrylonitrile (CTBN) rubber. The rubber was added as a prereacted concentrate containing 50% CTBN rubber and 50% epoxy resin. Two different concentrates were used. For the nonbrominated formulation, the concentrate was prepared by reacting 50% CTBN rubber with TGDDM. For the brominated formulation, a prereacted concentrate of the brominated diglycidyl type with 50% Br was used. Both concentrates were synthesized by following recommended procedures (2). The chemical structures for the various resins are shown below.

1. TRIS-(HYDROXYPHENYL) METHANE TRIGLYCIDYL ETHER

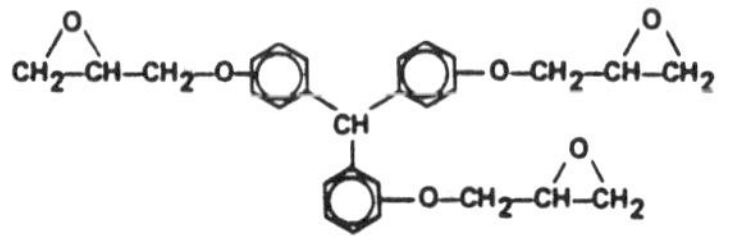

2. DIAMINO DIPHENYL SULFONE (DDS)

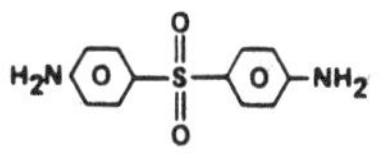

3. TETRAGLYCIDYL 4,4'-DIAMINO DIPHENYLMETHANE (TGDDM)

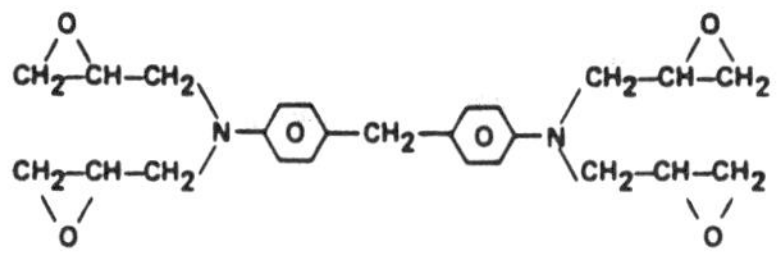

4. CARBOXYL-TERMINATED, LIQUID COPOLYMER OF BUTADIENE AND ACRYLONITRILE (CTBN)

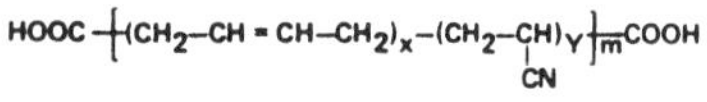

5. BROMINATED POLYMERIC ADDITIVE (BPA)

6. BPA-CTBN PREREACTED CONCENTRATE

Materials Used: The graphite cloth used for reinforcement was a product of Hexcel: F3T-584-42, 300, 3K 8 Harness satin weave with an epoxy resin sizing. The resins used include the following: TEN, an experimental resin of Dow Chemical Co. called XD 7342.00L (6) with an epoxide equivalent weight (EEW) of 162; F2001P, a brominated bisphenol epoxy resin from Makhteshim Chemical Works, Israel, with an EEW of 545 and 50% Br; TGDDM, a commercial product of Ciba-Geigy (MY-720); CTBN rubber, a commercial product of B. F. Goodrich, called Hycar 1300x13; and diamino diphenyl sulfone (DDS), an epoxy hardener with a stoichiometric ratio of 80%. The table shows the components of the various resins used.

Modification of Epoxy Resin with CTBN Rubber: The epoxy resin and the CTBN rubber were placed in a resin kettle and heated to 353 K; mixing was effective at that temperature. Triphenylphosphine (0.15%) was added under N_2 atmosphere. The reaction continued for 2 hrs. at temperatures of 403 K to 423 K. The reaction advancement was monitored by 0.1 N KOH/EtOH titration to determine the equivalent per hundred grams (EPHR) of the carboxyl group. The reaction continued to 1% of the initial EPHR.

FORMULATIONS OF COMPOSITES

COMPONENT	TGDDM	TGDDM WITH RUBBER	TEN	TEN WITH RUBBER	TEN WITH Br	TEN WITH Br RUBBER
TETRAFUNCTIONAL EPOXY RESIN	x	x				
TRIFUNCTIONAL EPOXY			x	x	x	x
EPOXY-RUBBER* (50/50 PREREACTED)		x		x		
F2001P (BPA WITH 50% Br)					x	x
F2001P-CTBN (50/50 PREREACTED)						x
DIAMINODIPHENYL SULFONE – DDS	x	x	x	x	x	x
METHYLETHYLKETONE (MEK, ml)	x	x	x	x	x	x

*50/50 PREREACTED CTBN WITH TGDDM.

The LOI and the Tg are given below. The LOI for the modified TGDDM and TEN epoxy resins show a decrease (higher flammability) with an increase in rubber content, a result of the more aliphatic character of the rubber. Composites prepared with brominated resin had higher LOI values than the nonbrominated composites. The LOI of rubber-modified TEN epoxy resin composites containing no bromine was 48.9%, with 19% Br the LOI was 69.3%, and with 38% Br the LOI was 86.6%. The composites with Br but no rubber gave even higher LOI values.

The modification of TGDDM and TEN epoxy resins with rubber and brominated resin does not have a deleterious effect on the Tg of the composites. This has been seen by other investigators (7) when working in the range of 5-10% rubber modification. The explanation of the increase in Tg is beyond the scope of this investigation. Since other investigators (8) have shown the additive effect of modification of epoxy resin, this area should be investigated in the future.

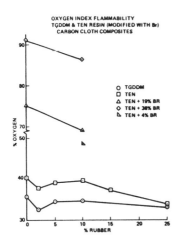

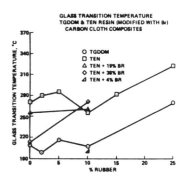

With two exceptions, the room-temperature flexural strengths of composites made with TGDDM and TEN epoxy resins increased with the addition of rubber. The flexural modulus was also closely retained or increased with the addition of rubber. Even with 25% rubber, the TGDDM and TEN epoxy resins retained 85% and 91%, respectively, of their original flexural modulus.

Samples modified with rubber and measured under hot/wet conditions showed more loss in flexural strength than did the TGDDM and TEN epoxy resin controls without rubber. However, the brominated and nonbrominated TEN epoxy resins were consistently higher than the modified TGDDM resin. For example, the TGDD resin with 10% rubber retained 84% and the brominated resin (19% Br) retained 62% of the original flexural strength when tested in hot/wet conditions.

The flexural modulus of the rubber-modified brominated and nonbrominated TEN epoxy resins measured in hot/wet conditions were uniformly higher than the modified TGDDM composites. In fact, they were equal to or better than the TGDDM control resin. For example, the hot/wet modulus of the TGDDM control resin was 54.8 GPa (96% retention) and the hot/wet TEN epoxy resin with 10% rubber was 59.2 GPa (108% retention).

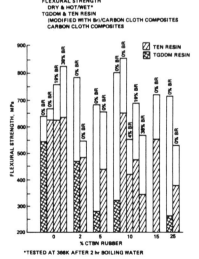

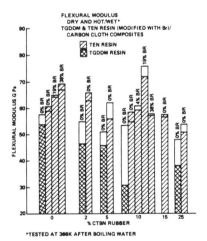

The maximum shear strength retention after impact was obtained on components made with 10 percent rubber modification. This was true for both the TEN and the TGDDM matrix resins as well as the two brominated TEN resins.

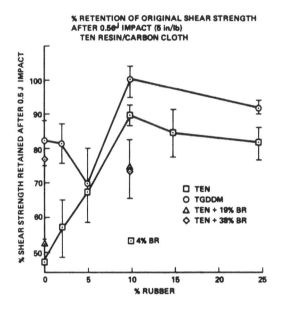

Instrumented Impact Tests: The instrumented impact testing was performed using two methods. The first method (A) used the General Research (formerly, Effects Technology) Dynatup Impact Tester (Model 8200). In this procedure a 16 mm hemispherical head impacts a test specimen at a constant velocity. This instrumented impact technique provides a complete record of the impact event. Automated data analysis provides a record of the applied load and energy absorbed during impact. The second method (B) used a Rheometrics Impact Tester. This method also uses a microprocessor to collect, calculate, and display impact data. In both cases the test specimen is supported by a metal frame. In method A the specimen is mounted horizontally, whereas in method B the specimen is mounted vertically. The use of both methods for testing the impact resistance of composites is well documented in the literature (7, 9, 10).

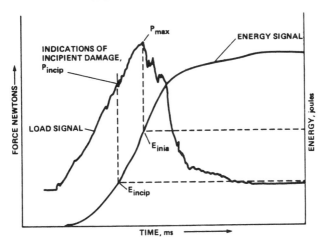

Instrumented Impact: The data for instrumented impact are given below. This test was performed only with composites made with the TEN epoxy resin. No instrumented impact was run on the TGDDM composites because of their low hot/wet flexural test values. The instrumented impact test specimens in all cases were 10.2 cm x 10.2 cm and thickness varied from 23 mm to 28 mm. The impact tests were run at room temperature.

Two values were reported for each test method. First is the maximum force (expressed in newtons) necessary for penetration of the test specimen. This is characterized as the delamination and fiber breaking of the composite specimen. The second value is the total energy absorbed during the impact event (expressed in J/cm thickness). The impact velocity of method A was 147 mm/sec and that of method B was 102 mm/sec. The maximum-force and energy-absorbed test results were higher for method B than for method A. However, the relative rankings of the samples by the two methods were, in general, in close agreement.

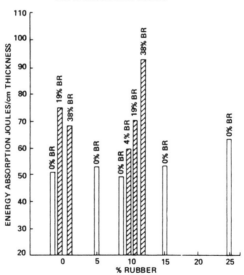

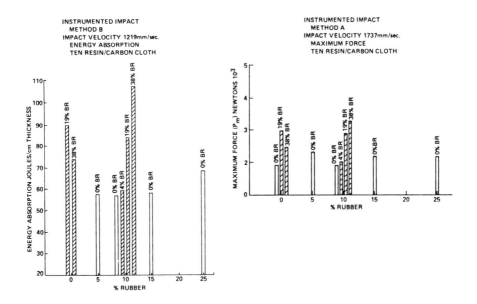
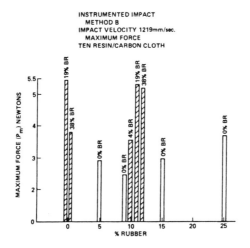

CONCLUSIONS

- MODIFICATION OF TEN RESIN WITH BROMINE GIVES BETTER IMPACT RESISTANCE THAN RUBBER MODIFICATION ALONE.

- 25% RUBBER ADDITION IS NECESSARY TO OBTAIN SIGNIFICANT IMPROVEMENT IN IMPACT RESISTANCE.

- IMPACT RESISTANCE INCREASES WITH BROMINE CONTENT.

- IMPACT VELOCITY DOES NOT SIGNIFICANTLY AFFECT THE ENERGY ABSORBED BY THE TEST SAMPLE.

- T_g DID NOT DECLINE WITH RUBBER MODIFICATION.

- TEN RESIN HAD BETTER HOT/WET PROPERTIES THAN TGDDM RESIN.

REFERENCES

1. D. A. Kourtides, Proceedings of the 21st AIAA Structures, Structural Dynamics and Materials Conference, May 12-14, 1980.

2. H. Drake and R. A. Siebert, SAMPE Quarterly, 6, No. 4 (1975).

3. E. H. Rowe and R. A. Siebert, Advances in Chemistry Series 154, ACS, Washington, DC (1976), pp. 326-346.

4. B. L. Lee, C. M. Lizak, C. K. Riew, and R. J. Moulton, Proceedings of the 12th National SAMPE Technical Conference, Seattle, Washington, Oct. 7-9, 1980.

5. Z. Nir, W. J. Gilwee, D. A. Kourtides, and J. A. Parker, Proceedings of the 41st Annual Technical Meeting (ANTEC) of the SPE, Chicago, Illinois, May 2-5, 1983.

6. K. L. Hawthorne, F. C. Henson, and R. Pinzelli, Organic Coatings and Applied Polymer Science Proceedings, ACS, 46, Las Vegas, Nevada, March 25-April 2, 1982, pp. 493-497.

7. F. Harper-Trevet, Proceedings of the 27th National SAMPE Symposium, San Diego, California, May 4-6, 1982.

8. G. Lee and B. Hartman, Journal of Applied Polymer Science, 28, 823-830 (1983).

9. G. F. Sykes and D. M. Stoakley, Proceedings of the 12th National SAMPE Technical Conference, Seattle, Washington, Oct. 7-9, 1980.

10. A. G. Miller, P. E. Hertzberg, and V. W. Rantala, Proceedings of the 12th National SAMPE Technical Conference, Seattle, Washington, Oct. 7-9, 1980.

MORPHOLOGY AND DYNAMIC MECHANICAL PROPERTIES OF DIGLYCIDYL ETHER OF BISPHENOL-A TOUGHENED WITH CARBOXYL-TERMINATED BUTADIENE-ACRYLONITRILE

Su-Don Hong, Shirley Y. Chung, Robert F. Fedors,
Jovan Moacanin and Amitava Gupta

*Jet Propulsion Laboratory
California Institute of Technology
Pasadena, California*

INTRODUCTION

The incorporation of a carboxyl-terminated butadiene acrylonitrile (CTBN) elastomer in diglycidyl ether bisphenol A (DGEBA) resin was reported to produce more than a 10-fold increase in the fracture toughness of the unmodified resin (1-9). The fracture toughness of fiber-reinforced composites containing such modified resins, however, has not always been reported to be increased (10-11). It was thought that these diverse results were at least in part due to the fact that the CTBN-modified DGEBA matrix had differing rubber particle size as well as size distribution which influenced the shape of the crack tip deformation zone (12) and hence the fracture toughness of the material. The particle sizes and size distribution of the rubber inclusions in a CTBN-modified epoxy can be affected by both the curing conditions and the chemistry and composition of the starting resin mixture (3,4,7,8,13,14). The presence of graphite fibers in the composite may further influence the morphology of the CTBN inclusions. Depending on the fiber surface treatment, the surfaces of graphite fibers may have reactive chemical groups (15,16) which will influence the cure kinetics of the resin and, consequently, possible change of the morphology of CTBN inclusions. Measurements of dynamic mechanical properties, scanning election microscopy and small-angle x-ray scattering were carried out to characterize the state of cure, morphology and particle size and size distribution of the neat resins and their graphite fiber reinforced composites, as summarized in Figure 1.

CHARACTERIZATION OF STATE OF CURE, MORPHOLOGY AND PARTICLE SIZES AND PHASE SEPARATION

- DYNAMIC MECHANICAL PROPERTIES TESTING
  - STATE OF CURE

- SCANNING ELECTRON MICROSCOPY
  - MORPHOLOGY OF FRACTURED SURFACES
  - DOMAIN SIZES FROM SEVERAL HUNDRED ANGSTROMS UP TO MICRONS

- SMALL-ANGLE X-RAY SCATTERING
  - DOMAIN SIZES IN THE RANGE OF 50Å TO 5000Å
  - PHASE SEPARATION

Figure 1

COMPOSITIONS OF TESTING SPECIMEN

Table 1 summarizes the compositions of the unmodified DGEBA base epoxy resin, trade name Hexcel 205 (HX-205), and the CTBN-modified epoxy resin, trade name F-185. Two fiber-reinforced composites, designated GD-31 and GD-48, made from Celion 6000 graphite fiber were also used for the testing. The matrix corresponding to GD-31 is F-185 and that for GD-48 is HX-205. Both composites are 6-ply laminates with unidirectional fiber layup. The resin content in both composites is about 37% by weight. All neat resins were supplied by Hexcel Corporation; composites were made and supplied by NASA Langley Research Center.

TABLE 1. COMPOSITIONS OF HX-205 AND F-185 RESINS

HX-205		F-185	
COMPONENT	APPROXIMATE WEIGHT %	COMPONENT	APPROXIMATE WEIGHT %
EPOXIDES			
(DIGLYCIDYL ETHER OF BISPHENOL A)			
(EPOXIDIZED NOVOLAC, EPOX. EQ. WT 165)	73	HX-205	86.5
DIPHENOLS			
(BISPHENOL-A)			
(TETRABROMOBISPHENOL-A)	20	Hycar 1300 x 9	8.1
CATALYST			
(DICYANDIAMIDE)			
(SUBSTITUTED UREA)	7	Hycar 1472	5.4

CHARACTERIZATION OF THE STATE OF CURE OF NEAT RESINS

The comparison of the tanδ versus temperature plots for HX-205 and F-185 are shown in Figure 2. F-185 shows an enhanced tanδ at temperatures above -50°C; at temperatures higher than about 25°C, tanδ for F-185 starts to increase drastically even though HX-205 and F-185 appear to have the same glass transition temperature. The glass transition temperatures of CTBN components are: Hycar 1300x9,-49°C and Hycar 1472,-24°C, as reported by the manufacturer (A. Siehert, private communication). F-185 does not show the separate transition peaks corresponding individually to Hycar 1300x9 and Hycar 1472. It appears that the CTBN components in F-185 do not form a pure rubber phase; instead, the rubber phase is blended with the epoxy resin to form CTBN-rich domains.

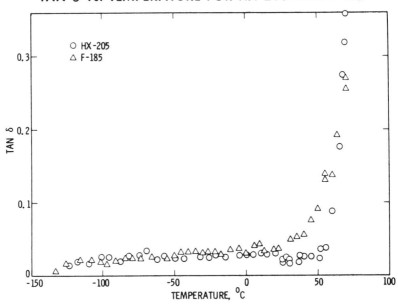

Figure 2

CHARACTERIZATION OF THE STATE OF CURE OF FIBER-REINFORCED COMPOSITES

The plots of tan δ as a function of temperature for the composites GD-48 and GD-31, shown in Figure 3, are very similar to the corresponding plots for HX-205 and F-185 neat resins shown in Figure 2. Thus the dynamic mechanical property characterization indicates that HX-205 both as neat resin and matrix material in GD-48 composite and F-185 as neat resin and matrix in GD-31 composite have similar states of cure. For both F-185 neat resin and matrix, extensive mixing of CTBN rubber and epoxy resin occurs.

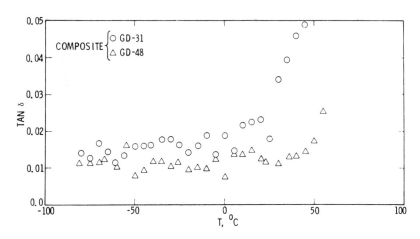

Figure 3

MORPHOLOGY OF FRACTURE SURFACES OF NEAT RESINS

The scanning electron micrographs of fracture surfaces of both HX-205 and F-185 neat resins are shown in Figure 4. The fracture surface of HX-205 is very smooth, indicative of typical brittle fracture behavior. On the other hand, F-185 has a very rough fracture surface, indicating that the resin was highly strained before fracture occurred. There are also some craters which appear to represent the separation of spheroidal rubber domains from the matrix.

SEM MICROGRAPHS OF FRACTURE SURFACES OF HX-205 AND F-185 RESINS

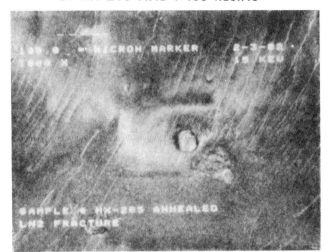

HX-205

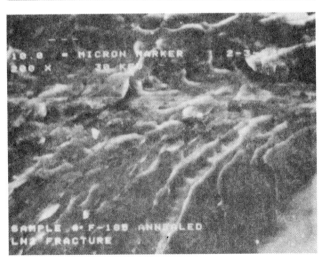

F-185

Figure 4

Matrix Synthesis and Characterization

MORPHOLOGY OF FRACTURE SURFACES OF FIBER-REINFORCED COMPOSITES

The scanning electron micrographs of GD-31 (F-185 / graphite fiber composite) and GD-48 (HX-205/graphite fiber composite) are shown in Figure 5. The GD-48 laminate gave a relatively clean fracture with no sign of the resin being strained before fracture occurred. On the other hand, the GD-31 laminate exhibited a very rough fracture surface with indications that some regions of the matrix were highly strained before fracture.

HX-205/GRAPHITE FIBER COMPOSITE
(GD-48)

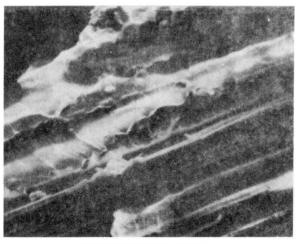

F-185/GRAPHITE FIBER COMPOSITE
(GD-31)

Figure 5

SAXS CHARACTERIZATION OF PARTICLE SIZE AND SIZE DISTRIBUTION

Small angle x-ray scattering was carried out to determine if the morphologies of the CTBN-rich domains in the neat resin and in the composite matrix are similar. Results for both HX-205 and F-185 neat resins as well as their corresponding composites are shown in Figure 6. In the scattering angle range 0.7×10^{-3} to 40×10^{-3} radians, the F-185 neat resin has a higher scattering intensity, by a factor of about 10 in the lower angle region, than does the HX-205 neat resin. This indicates that there are smaller rubber-rich domains having sizes of the order of 100Å to several thousand angstroms present in the CTBN-toughened neat resin. A comparison of the scattering profiles for both the F-185 neat resin and the GD-31 composite indicates that both have nearly identical scattering intensities at scattering angles lower than 4×10^{-3} radians. The scattering intensity at larger angles for F-185 in the composite is much higher than that for the F-185 neat resin. Analysis of the data[17] indicates that there is a larger fraction of smaller CTBN domains in the F-185 matrix of the composite than in the F-185 neat resin.

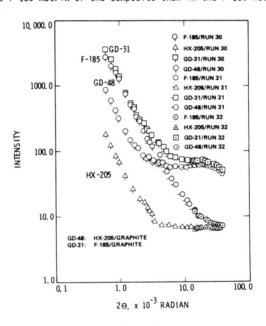

Figure 6

CONCLUSION

The HX-205 and F-185 neat resins and the corresponding composites appear to have the same state of cure as characterized by dynamic mechanical properties. The CTBN-rich domains in F-185 neat resin and F-185 composite resin have sizes ranging from 50 Å or smaller to 40 μm and larger. The F-185 material, both as neat resin and as matrix, shows a ductile fracture behavior, indicating a toughening effect due to incorporation of CTBN rubber. The morphology of the CTBN domains in the F-185 matrix, as determined by small-angle X-ray scattering, appears to be different from that in the neat resin. There is a larger fraction of smaller sizes of CTBN domains existing in the F-185 matrix as compared to the corresponding F-185 neat resin. Because CTBN domains in the size range of the order of several hundred angstroms are less effective in increasing fracture toughness, this fact may partially explain the reported observation that some composites made with the CTBN-modified DGEBA epoxy resins did not show significant improvement in fracture toughness. It is emphasized that the neat resin and the corresponding matrix prepared from the identical resin material may not have similar morphology even when prepared using the same curing procedure.

REFERENCES

1. W. D. Bascom, R. J. Moulton, E. H. Rowe and A. R. Siebert, Org. Coat. Plast., Preprint, 1978, 39, 164.
2. F. J. McGarry, Proc. Roy. Soc. London, 1970, A319, p. 59.
3. E. H. Rowe, A. R. Siebert and R. S. Drake, Mod. Plast., 1970, 417, 110.
4. J. N. Sultan, R. C. Laible and F. J. McGarry, Appl. Polymer Symp., 1971, 16, 127.
5. J. N. Sultan and F. J. McGarry, Polym. Eng. Sci., 1973, 13, 29.
6. W. D. Bascom, R. L. Cottington, R. L. Jones and P. Peyser, J. Appl. Polymer Sci., 1975, 19, 2545.
7. C. K. Riew, E. H. Rowe, and A. R. Siebert, ACS ADVANCES IN CHEMISTRY, 1976, SERIES No. 154, p. 326.
8. C. B. Bucknall and T. Yoshii, Brit. Polym. J., 1978, 10, 53.
9. W. D. Bascom and D. L. Hunston, Plastic and Rubber Institute, London, Preprints, 1978, 1, p.22.
10. G. B. McKenna, J. F. Mendell and F. J. McGarry, Soc. Plastic Industry, Ann. Tech. Conf 1974, Section 13-C.
11. J. M. Scott and D. C. Phillips, J. Mat. Sci., 1975, 10, 551.
12. W. D. Bascom, J. L. Bitner, R. J. Moulton and A. R. Siebert, Composites, January 1980, 9.
13. A. C. Meeks, Polymer, 1974, 15, 675.
14. T. T. Wang and H. M. Zupko, J. Appl. Polym. Sci., 1981, 26, 2391.
15. F. Hopfgarten, Fiber Sci. Technol., 1978, 11, 67.
16. G. E. Hammer and L. T. Drzal, Applications of Surface Science, 1980, 4, 340.
17. S. D. Hong, S. Y. Chung, G. Neilson and R. F. Fedors, in "Characterization of Highly Cross-linked Polymers." ASC Symposium series 243, page 91-108, Ed. SS. Labana and R. A. Dickie, 1984.

MATRIX RESIN CHARACTERIZATION IN CURED GRAPHITE COMPOSITES USING DIFFUSE REFLECTANCE-FTIR

Philip R. Young
NASA Langley Research Center
Hampton, VA

A.C. Chang
Kentron Technical Center
Hampton, VA

INTRODUCTION

The chemical characterization of cured graphite fiber reinforced polymer matrix composites is complicated by the fact that the resins are insoluble and the composites are opaque. Standard analyses which depend either on the ability to dissolve the sample or to detect transmitted radiation are impossible. As a result, data reported on environmentally exposed composites primarily concerns macroscopic information such as weight loss or changes in selected mechanical properties. The correlation of changes in resin molecular structure with this information could lead to a more fundamental understanding of why composite performance generally deteriorates with environmental aging.

The objective of the present research is to gain a basic chemical understanding of composite and adhesive behavior. Our approach has been to develop diffuse reflectance in combination with Fourier transform infrared spectroscopy to gain access to this information. Several composite and adhesive materials were characterized before and after environmental exposure. In each case significant changes in resin molecular structure were observed and correlated with changes in mechanical properties, providing new insights into material performance.

RESIN CHARACTERIZATION IN CURED GRAPHITE FIBER REINFORCED COMPOSITES USING DIFFUSE REFLECTANCE-FTIR

OBJECTIVE: GAIN A FUNDAMENTAL CHEMICAL UNDERSTANDING OF CURED COMPOSITE MATERIALS AND POLYMERIC ADHESIVES

APPROACH: DEVELOP DIFFUSE REFLECTANCE-FTIR AS AN ANALYTICAL TECHNIQUE TO OBTAIN MOLECULAR LEVEL INFORMATION ON CURE AND ENVIRONMENTAL BEHAVIOR

REFLECTANCE TECHNIQUES

Diffuse reflectance infrared spectroscopy using dispersive monochromators has been used for a number of years to obtain optical information about opaque materials (ref. 1). However, the spectra obtained were not generally considered to be high quality, particularly where organic compounds were involved. Recent work (refs. 2 through 4) has shown that the combination of diffuse reflectance with Fourier transform infrared spectroscopy (DR-FTIR) can lead to high-quality spectra of a variety of samples. Our contribution has been to develop and demonstrate the applicability of the technique for high-performance composites and adhesives.

Reflected radiation is composed of specular and diffuse components. Specular reflection is mirror-like in that the angle of incidence is equal to the angle of reflection. Since selected wavelengths are absorbed upon surface reflection, this component contains optical information about the sample. Composites are not good specular reflectors in the infrared. However, they do exhibit a significant diffuse component.

Diffuse reflection arises from radiation penetrating into the interior of a sample and undergoing multiple reflections before reemerging. Selected wavelengths are again absorbed at each reflection. Thus, the diffuse component also contains valuable optical information.

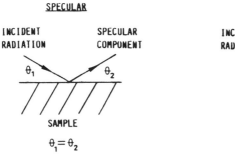

DIFFUSE REFLECTANCE OPTICS

A schematic of the commercially available optics used to make this study is given below. The unit sits inside the sample compartment of the FTIR optical bench. Radiation from the interferometer eventually reflects from an off-axis ellipsoidal mirror onto the sample which is placed in a holder and adjusted to sit at the focal point of the mirror. A second ellipsoidal mirror collimates the diffusely reflected components and passes them to the detector.

Spectra may be obtained by reflecting radiation directly off the surface of a small (0.8 cm x 0.8 cm) piece of composite placed in the sample mount. More definitive spectra are usually obtained by filing into the composite and mixing the resulting powder at 3% by weight with potassium bromide (KBr). These two sampling techniques are discussed later in greater detail.

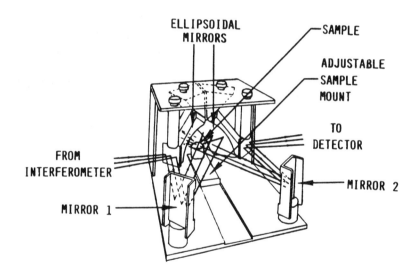

DR-FTIR SPECTRUM OF CURED GRAPHITE COMPOSITE

This figure shows the spectrum of a fully cured graphite/polyimide matrix resin composite and serves as an example of the quality of spectra that can be obtained. This spectrum, of a powdered sample mixed with KBr, is easily interpretable. The ordinate gives the percent reflectance, the ratio of the sample reflectance to that of a powdered KBr reference. This axis is converted to absorbance in spectra presented subsequently. Generally, 10 to 40 percent of the total energy available is retained by DR-FTIR. Thus, there is no need to expand very weak signals to yield an apparent enhanced spectrum. Typically, 512 scans are taken at 1.8 seconds per scan and 4 wavenumber resolution. Vendor-supplied software was quite adequate for mathematical computations and data manipulation.

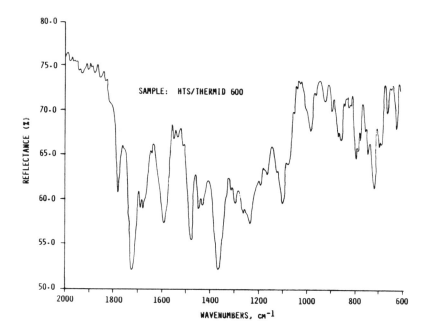

DR-FTIR SPECTRUM OF CELION 6000 GRAPHITE FIBER

Several experimental aspects of DR-FTIR will be discussed before considering initial applications of the technique. The figure below gives the spectrum of Celion 6000 graphite fiber. A fairly flat baseline was obtained. The minor peak around 2400 wavenumbers (cm^{-1}) is due to carbon dioxide which was not completely purged from the sample compartment. Since no significant peaks are attributable to the graphite fiber, any peaks observed for a composite are assumed to be due only to the matrix resin. Attempts to record a spectrum of the sizing on commercial fibers have been unsuccessful.

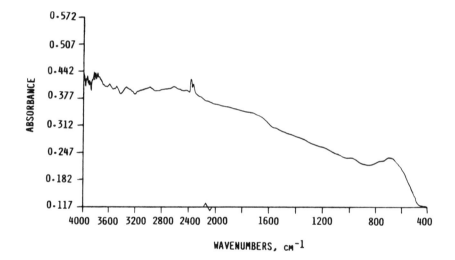

COMPARISON OF SPECTRA OBTAINED BY DIFFUSE REFLECTANCE AND KBr TRANSMISSION

Diffuse reflectance can produce artifacts not normally encountered with transmission spectroscopy. The spectra below were determined on an uncured matrix resin with no graphite fiber present. The solid-line DR-FTIR spectrum was generated by mixing the resin at 4% by weight with KBr. The dashed-line spectrum was obtained by standard transmission techniques after pressing the above resin/KBr mix into a pellet. Thus, the figure compares diffuse reflectance with transmission on essentially the same sample.

Diffuse reflectance tends to enhance weak bands and suppress strong ones. This enhancement may be an advantage, particularly if the weak bands have significance. In this figure, the band at 3241 cm^{-1} is most likely due to unreacted acetylene groups and could be used to monitor cure. Further analysis of the figure reveals that wavelengths of peaks obtained by reflectance and transmission are not always the same.

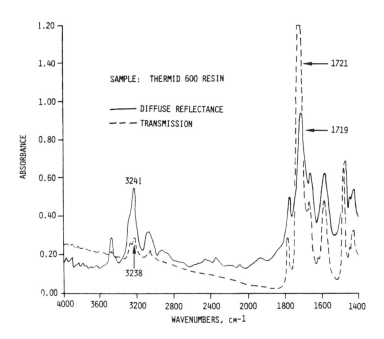

COMPARISON OF SPECTRA OBTAINED BY SURFACE REFLECTANCE AND POWDERED SAMPLE TECHNIQUES

Very strong bands may be distorted when spectra are determined using the composite surface rather than by mixing a small amount of powdered composite with KBr. This is illustrated in the figure for a cured polyimide matrix resin composite. The spectra are separated for clarity and absorbance values are not given on the ordinate.

A normal appearing imide carbonyl at 1718 cm^{-1} is obtained by the powdered sample/KBr mix technique. This carbonyl is distorted when the unaltered composite surface is analyzed directly. This artifact initially created confusion and almost led to the erroneous conclusion that the chemistry of the resin on the surface of the composite was different from that beneath the surface. This behavior has since been identified as due to interference by a specular component in regions where distortion occurs (ref. 5).

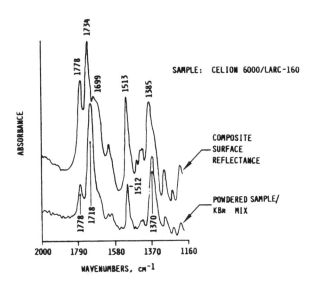

COMPARISON OF INFRARED SAMPLING TECHNIQUES

The next two figures compare diffuse reflectance spectra with spectra obtained by more conventional techniques. Diffuse reflectance is compared with multiple internal reflectance or attenuated total reflectance (ATR) in this figure, where spectra for a polyimide matrix resin composite are given. As previously discussed, the imide carbonyl around 1700 cm^{-1} is distorted by DR-FTIR. However, the diffuse spectrum obviously contains more information. This laboratory has not had success obtaining quality spectra of cured composites by ATR.

As a sampling note, the top spectrum was recorded on a 0.8 cm x 0.8 cm piece of composite. ATR generally requires a much larger sample. In addition, the sample surface must be smooth to provide good contact and avoid scratching the ATR optical crystal.

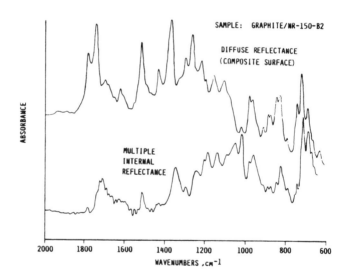

COMPARISON OF INFRARED SAMPLING TECHNIQUES (CONTINUED)

Diffuse reflectance and KBr transmission spectra are compared in this figure for the same polyimide matrix resin composite analyzed in the preceding figure. transmission spectrum, the best of several attempts, contained more information than anticipated. If enough FTIR scans are made and the data are optimized, surprisingly good KBr pellet spectra of these materials can be generated. However, the transmission of KBr increases around 1000 cm^{-1}. This results in a baseline inflection, a problem not encountered with diffuse reflectance where the baseline remains relatively flat. The reflectance spectrum is deemed to be superior since it is more informative.

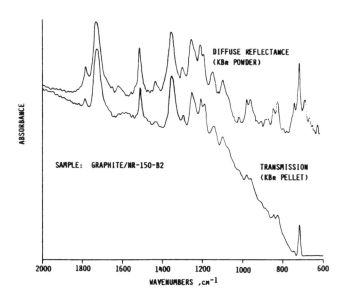

DIFFUSE REFLECTANCE ARTIFACTS

This figure summarizes several experimental aspects of diffuse reflectance spectroscopy. Although supportive spectra are not presented in this report, particle size can also affect the quality of spectra when analyzing by the powdered sample technique. The procedure used in this research of filing into the composite with a jeweler's file provided an approximately constant 100 μm particle size.

- ENHANCEMENT OF WEAK BANDS
- SUPPRESSION OF STRONG BANDS
- λ BY % R MAY DIFFER FROM λ BY % T
- PARTICLE SIZE MAY AFFECT SPECTRUM QUALITY (POWDERED SAMPLE TECHNIQUE)
- POSSIBLE DISTORTION OF STRONG BANDS (SPECIMEN REFLECTANCE)

INITIAL APPLICATIONS OF DR-FTIR

Several different materials have been examined by DR-FTIR to demonstrate the usefulness of the technique for the analysis of high-performance composites and adhesives. These materials are summarized below. Whenever possible, samples were failed mechanical test specimens which other researchers had exposed to various environments, tested, and then reported. Details of these earlier studies are found in references 6-9.

Note that the reported property changes are macroscopic in nature, such as a change in tensile strength or a weight loss. The ultimate goal of the present research is to provide a molecular level understanding of these changes. This level of information has not been available in the past due to the intractable nature of the materials undergoing analysis.

MATERIAL	ENVIRONMENTAL EFFECT	SIGNIFICANT PROPERTY CHANGE
GR/POLYSULFONE: CELION 6000/P-1700	ELECTRON RADIATION 10^{10} RAD	25° INCREASE IN T_G 24% DECREASE IN OXYGEN [1.] CONTENT
GR/EPOXY: AS/3501-5	THERMAL AGING 250°F/50,000 HRS	NO DECREASE IN TENSILE STRENGTH
GR/POLYIMIDE: A. HTS/SKYBOND 710	THERMAL AGING 450°F/25,000 HRS 550°F/25,000 HRS	NO CHANGE IN TENSILE STRENGTH AT 450°F; 80% DECREASE AT 550°F
B. CELION 6000/LARC-160	THERMAL AGING 450°F/15,000 HRS	13° INCREASE IN T_G 8% WEIGHT LOSS
ADHESIVE: EXPTL. POLYIMIDE-SULFONE	CURE	IMIDIZATION LOSS OF VOLATILES

1. THIN-FILM DATA

DR-FTIR SPECTRA OF GRAPHITE/POLYSULFONE COMPOSITE
BEFORE AND AFTER RADIATION EXPOSURE

This figure shows a portion of the spectrum of a graphite/polysulfone composite before and after receiving 10^{10} rad of electron radiation. The polymer repeat unit is included in the figure. Degradation is apparent; bands associated with the $-SiO_2-$group (1409, 1294, and 1150 cm^{-1}) have disappeared or decreased in intensity as the result of exposure. As noted in the preceding figure, a loss of oxygen was reported for polysulfone films irradiated under the same conditions (ref. 6). Apparently, additional cross-linking also occurred to increase the Tg of this material. The aromatic rings do not appear to be affected by radiation since no decrease in intensity or change in band shape is observed in the spectrum for those groups.

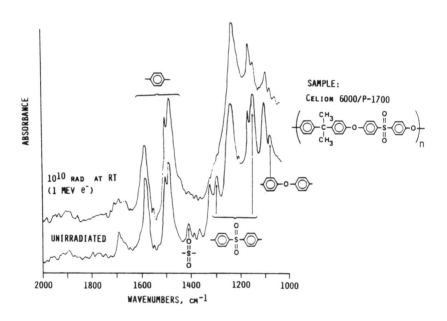

DR-FTIR SPECTRA OF GRAPHITE/EPOXY COMPOSITE
BEFORE AND AFTER THERMAL AGING

Spectra of graphite/epoxy composites thermally aged at 250°F are given below along with monomer unit structures. Phenyl bands (1594 and 1512 cm^{-1}) and sulfone bands (1389, 1287, and 1143 cm^{-1}) have not changed significantly, indicating that these portions of the polymer backbone are not adversely affected by extended exposure at 250°F. However, new bands did develop between 1700 and 1650 cm^{-1}. This is apparently due to $>$C=O which resulted from the oxidation of -CH$_2$- or residual -OH groups in the epoxy resin. This oxidation was not reflected in longitudinal tensile strength measurements (refs. 7, 8) for these specimens. Spectra of composites aged at 350°F showed extensive oxidation to $>$C=O above 1700 cm^{-1}. A decrease in tensile strength reported for specimens aged at this temperature may reflect a loss of resin in the aliphatic portions of the matrix resin.

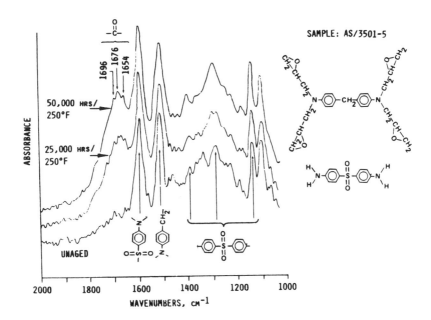

DR-FTIR SPECTRA OF GRAPHITE/POLYIMIDE COMPOSITE
BEFORE AND AFTER THERMAL AGING

This figure gives spectra of a thermally aged linear polyimide matrix resin composite. No shift in frequency or change in band intensity is noted after 25,000 hours at 450°F. No change in tensile strength was reported for this exposure (refs. 7, 8). However, band intensity had decreased markedly after 25,000 hours at 550°F, suggesting resin loss. A very intense band centered at 1105 cm^{-1} is apparent after 25,000 hours at the higher temperature. This was initially thought to be due to the formation of carbon-oxygen ether linkages. However, further study proved this to be due to silicon-oxygen vibrations.

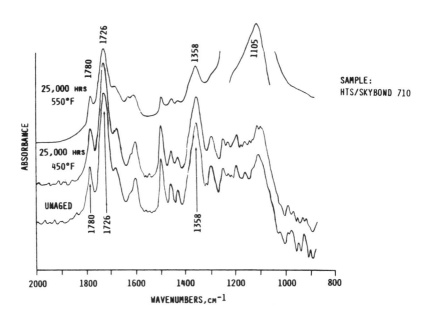

SPECTRA OF THERMALLY AGED SKYBOND 710 COMPOSITE AND CAB-O-SIL

The spectrum of the residue remaining after 50,000 hours at 550°F is given below. The spectrum of CAB-O-SIL is included in the figure, leaving little doubt as to the identity of the residue observed in the thermally aged composite. Apparently, a vendor added CAB-O-SIL to the resin to adjust the viscosity during the prepregging operation. This event did not become apparent until after the material had been processed into composites and then aged for a considerable period of time.

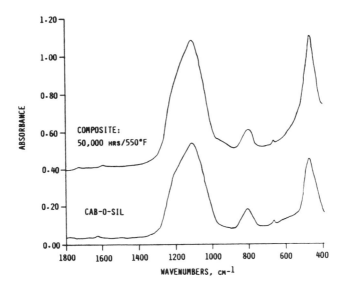

DR-FTIR SPECTRA OF GRAPHITE/ADDITION POLYIMIDE COMPOSITE
BEFORE AND AFTER THERMAL AGING

An addition polyimide matrix resin composite showed significant changes as the result of thermal aging. New bands at 1667 and 931 cm^{-1} are apparent in the 15,000 hr/450°F spectrum. This is due to -CH$_2$-groups having been oxidized to $>$C=O groups. When this occurred, phenyl bands at 1599 and 1512 cm^{-1} switched in intensity due to conjugation with the new carbonyl group.

Shifts in the position of two bands associated with imide vibrations are also noted. Bands at 1716 and 1372 cm^{-1} for the unaged composite shifted to 1724 and 1361 cm^{-1}, respectively, after aging. These bands did not move for the linear polyimide matrix resin composite after aging for 25,000 hours at 550°F. A detailed DR-FTIR study involving both model compounds and sets of isothermally aged composites (ref. 9) strongly suggested that this movement was indicative of additional cure and/or crosslinking which constrained the vibration of the imide rings.

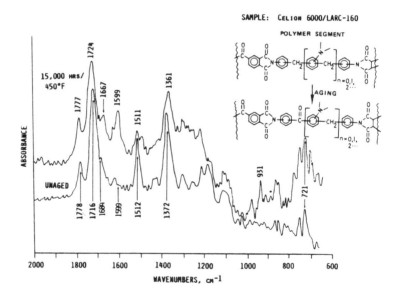

DR-FTIR SPECTRA OF POLYIMIDE-SULFONE ADHESIVE DURING CURE

DR-FTIR appears promising as a technique for studying the cure of adhesive systems. The figure shows spectra at various stages during the cure of an experimental polyimide-sulfone adhesive being developed at NASA Langley Research Center. The spectra were obtained on a piece of glass tape coated with the adhesive.

Bands due to amide and acid portions of the adhesive are apparent in the spectrum of the wet tape. Aliphatic solvent is also present. Progressive staging at 100°C and 160°C removed some solvent and slightly imidized the resin. Bonding appears to completely imidize the material. Bands associated with the pre-polymer are now gone and a strong imide carbonyl dominates the spectrum. Aromatic protons, previously hidden, are now apparent by the absorption slightly above 3000 cm^{-1}. A weak imide overtone appears around 3500 cm^{-1}.

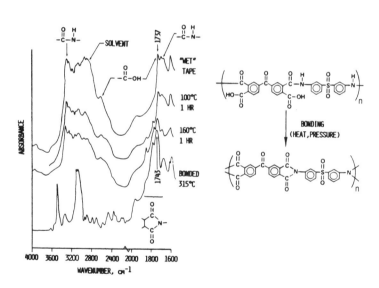

COMPARISON OF REGULAR PRESS AND RAPID BONDING OF POLYIMIDE-SULFONE ADHESIVE ON TITANIUM

NASA Langley is developing a rapid method for bonding joints. With this method, the specimen is taken through a two-minute cure cycle as opposed to the standard cycle of placing the specimen in a press, bringing the press to temperature, and applying pressure. The latter procedure can take up to two hours. The question was raised as to whether the resin chemistry was the same for the two cure cycles.

The figure shows spectra of failed titanium lap shear specimens bonded by the two methods. The polyimide-sulfone adhesive shown in the preceding figure was used for bonding. Small portions (0.8 cm x 0.8 cm) of the titanium adherend containing the adhesive were analyzed directly. The two spectra are virtually superimposable, leading to the conclusion that the resin chemistry has not been altered by the rapid bonding procedure.

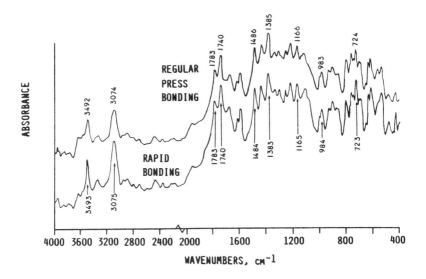

DR-FTIR SPECTRA OF TOUGHENED EPOXY COMPOSITE

DR-FTIR was also demonstrated to provide information on toughened materials. The top spectrum (C6000/HX205) was determined on a standard graphite/epoxy resin composite. The bottom spectrum (C6000/F185) is for the same epoxy resin modified by the addition of about 8% of a liquid carboxy-terminated butadiene-acrylonitrile (CTBN) and 5.4% of a solid CTBN.

The spectra are interpretable; phenyl bands (~1600, 1580, 1500, and 1450 cm^{-1}) and an ether band (~1240 cm^{-1}) are readily apparent. Both resins were cured with dicyanodiamide (DICY). Absorption around 2175 cm^{-1} is due to C $\equiv$ N in the curing agent. The small band at 2236 cm^{-1} for the toughened composition is apparently due to nitrile in the CTBN. The carboxy-terminated portion of the toughener can be observed using FTIR subtractive techniques.

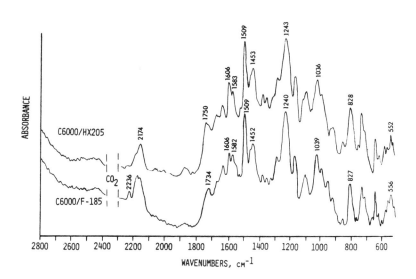

SUMMARY

The combination of diffuse reflectance with Fourier transform infrared spectroscopy has been demonstrated as an effective technique for gaining molecular level information on advanced materials. Several different graphite fiber reinforced polymeric matrix resin composites exposed to various environmental conditions were examined. In each case, significant changes in resin molecular structure were observed. These changes provided insights into previously reported changes in composite weight or mechanical properties. The technique was also shown to be applicable to adhesives and toughened systems. DR-FTIR is anticipated to play an increasingly important role in the characterization of advanced composites and adhesives, pointing towards new directions for the analysis of environmentally stable and processable polymers for efficient aerospace structures.

DIFFUSE REFLECTANCE-FTIR SPECTROSCOPY PROVIDES

- ACCESS TO PREVIOUSLY INACCESSIBLE MOLECULAR INFORMATION
- PRECISION AND REPEATABILITY
- FUNDAMENTAL INSIGHTS INTO COMPOSITE BEHAVIOR
- CONCEPTS FOR IMPROVED RESIN SYSTEMS

REFERENCES

1. Wendlant, W. W., editor: Modern Aspects of Reflectance Spectroscopy, Plenum Press, New York (1968).

2. Fuller, M. P. and Griffins, P. R.: Anal. Chem., 50, 1906 (1978).

3. Fuller, M. P. and Griffins, P. R.: Appl. Spectrosc., 34, 533 (1980).

4. Maulhardt, H. and Kunath, D.: Appl. Spectrosc., 34, 383 (1980).

5. Smyrl, N. R., Fuller, E. L., and Powell, G. L.: Appl. Spectrosc., 37, 38 (1983).

6. Santos, B. and Sykes, G. F.: Radiation Effects on Four Polysulfone Films, SAMPE Preprints, 13, 256 (1981).

7. Kerr, J. R. and Haskins, J. F.: Time-Temperature Stress Capabilities of Composite Materials for Advanced Supersonic Technology Applications. NASA CR-159267, April 1980.

8. Kerr, J. R. and Haskins, J. F.: Effects of 50,000 Hours of Thermal Aging on Graphite/Epoxy and Graphite/Polyimide Composites. Paper presented at the AIAA/ASME/ASCE/AHS 23rd Structures, Structural Dynamics, and Materials Conference, New Orleans, LA, May 1982.

9. Nelson, J. B.: Thermal Aging of Graphite/Polyimide Composites. Long-Term Behavior of Composites, ASTM STP 813, O'Brien, T. K., Ed., American Society for Testing and Materials, Philadelphia, 206 (1983).

SOLVENT RESISTANT THERMOPLASTIC COMPOSITE MATRICES

P.M. Hergenrother and B.J. Jensen
NASA Langley Research Center
Hampton, VA

S.J. Havens
Kentron International, Inc.
Hampton, VA

INTRODUCTION

More extensive use of resin matrix/fiber reinforced composites in commercial aircraft has been identified as a viable means of reducing the weight and thereby increasing the fuel efficiency and payload. Although major improvements in composite properties (e.g. better damage tolerance and moisture resistance) have been made by improving the resin matrix (e.g. rubber and thermoplastic toughened epoxies, epoxy/bismaleimide and cyanate/bismaleimide blends), these new systems still exhibit shortcomings such as limited prepreg shelflife, unforgiving and long cure cycles, and cocuring problems. Another approach to better resin matrices involves the modification of thermoplastics. Commercial thermoplastics such as UDEL® polysulfone and Victrex® polyethersulfone are excellent engineering materials with good toughness and thermoformability but poor solvent resistance. The latter feature is an important requirement in composite structures on commercial airplanes which demand ~50,000 hours of service at temperatures from $-54^{\circ}C$ to $93^{\circ}C$ in an environment which includes exposure to moisture and aircraft fluids (e.g. hydraulic fluid and paint strippers) while under load. The problem, objective and technical approach of this research are summarized in Figure 1.

Problem

Thermoplastics (e.g. polysulfones) as structural resins (e.g. adhesives and composite matrices) are sensitive to aircraft fluids (e.g. hydraulic fluid and paint stripper) and, upon exposure, undergo a loss of mechanical properties, especially in a stressed condition.

Objective

The objective is to develop new concepts and technology that improve the solvent resistance of thermoplastics without severely compromising the attractive features such as toughness and thermoformability.

Approach

The approach is to modify thermoplastics with ethynyl (acetylenic) groups which undergo thermally induced addition reactions (no volatiles) to improve the solvent resistance.

Figure 1

MODIFIED POLYSULFONES WITH IMPROVED SOLVENT RESISTANCE

Previous attempts to modify polysulfones to primarily improve the solvent resistance are listed in Figure 2. In the mid-1970's, Union Carbide Corporation introduced RADEL®, a sulfone polymer containing biphenyl moieties, which exhibited better solvent resistance than UDEL®. However, RADEL® was more difficult to process than UDEL® and still did not have the solvent resistance required for composite application on commercial aircraft. An experimental material from Union Carbide Corporation, PKXA (sulfone polymer end-capped with trimethoxysilyl groups which underwent hydrolysis and subsequent reaction to yield siloxane moieties), also failed to exhibit the required solvent resistance. Work at AFWAL involved the blending of a reactive plasticizer, bis[4-(3-ethynylphenoxy)phenyl]-sulfone, with UDEL® which, upon curing, exhibited improved solvent resistance (1). Sulfone block copolymers containing crystalline regions were reported to exhibit improved solvent resistance (2). Cured nadimide-terminated polysulfone, designated NTS-20, was initially reported to have good resistance to methylene chloride, exhibiting only slight swelling in composite form after 2 months immersion (3). Later work revealed severe loss of mechanical properties after 28 days exposure to methylene chloride (4). Cured ethynyl (acetylenic)-terminated sulfone oligomers exhibited good resistance to hydraulic fluid and chloroform when the linear sulfone segment was relatively short [number average molecular weight ($\bar{M}_n$) ~ 3000 g/mole] (5).

o RADEL® and PKXA, Union Carbide Corp., Mid-1970's

o Blend of reactive plasticizer with UDEL®, AFWAL, 1978

o Semi-Crystalline Sulfone Block Copolymers, VPI&SU, 1979

o Nadimide-Terminated Polysulfone (NTS-20), Boeing Aerospace Co., 1980

o Ethynyl-Terminated Sulfone Oligomers (ETS), NASA Langley, 1982

Figure 2

APPROACHES INVESTIGATED

Various approaches were investigated to improve the solvent resistance of thermoplastics and are listed in Figure 3. Ethynyl groups were placed on the ends of sulfone oligomers as discussed in subsequent figures. Ethynyl groups were incorporated pendent along the backbone of sulfone/ester polymers as presented later in this paper. Ester oligomers were also end-capped with ethynyl groups. To further investigate the effect of pendent ethynyl groups on the properties of cured resins, phenoxy resins containing pendent ethynyl groups were prepared and characterized. The last approach involved the use of a coreactant which was blended with a phenoxy resin containing pendent ethynyl groups.

o Ethynyl groups on the ends of sulfone oligomers

o Ethynyl groups pendent on sulfone/ester polymer

o Ethynyl groups on the ends of ester oligomers

o Ethynyl groups pendent on phenoxy resins

o Coreactant blended with phenoxy resins containing pendent ethynyl groups

Figure 3

SYNTHESIS OF HYDROXY-TERMINATED SULFONE OLIGOMERS

Hydroxy-terminated sulfone oligomers of different molecular weights were synthesized according to a known procedure (6) from the reaction of 2,2-bis(4-hydroxyphenyl)propane (bis-phenol A) and 4,4'-dichlorodiphenyl-sulfone using potassium carbonate in N,N-dimethylacetamide as depicted in Figure 4. The hydroxy-terminated sulfone oligomers were characterized by titration to determine hydroxyl end-groups (7) and accordingly $\bar{M}_n$, size exclusion chromatography, differential scanning calorimetry to obtain the glass transition temperature (Tg) and inherent viscosity as presented in table 1. Two hydroxy-terminated sulfone oligomers are of particular interest in this paper: one with a $\bar{M}_n$ up ~12000 g/mole employed in the synthesis of the ethynyl-terminated sulfone oligomer and the other with a $\bar{M}_n$ of ~4000 g/mole used to prepare the sulfone/ester polymer containing pendent ethynyl groups. Characterization data on these two oligomers are presented below.

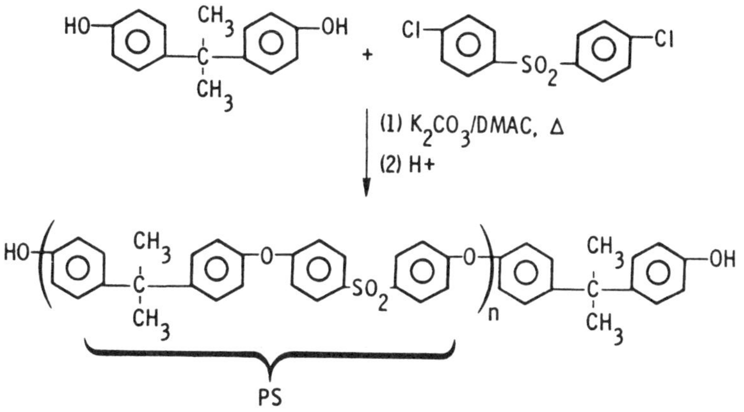

Figure 4

TABLE 1

$\bar{M}_n$ g/mole	η_{inh}, dl/g[1]	SEC Peak Retention Time, min[2]	Tg, °C[3]
4,000	0.09	18.90	167
12,000	0.38	17.65	183

[1] Inherent viscosity determined on 0.5% solution in chloroform at 25°C

[2] Size exclusion chromatography determined using a bank of μ-Styragel columns ($10^6, 10^5, 10^4$ and 10^3 A°) and chloroform as solvent

[3] Glass transition determined by differential scanning calorimetry (DSC) at a heating rate of 20°C/min

SYNTHESIS OF 4-ETHYNYLBENZOYL CHLORIDE

The 4-ethynylbenzoyl chloride was prepared as shown in Figure 5 according to a known procedure (8). The acid chloride was obtained as a yellow crystalline solid, melting point (mp), 75ºC to 76ºC, after recrystallization from hexane. A lower cost route to 4-ethynylbenzoyl chloride has since been developed which utilizes 2-methyl-3-butyn-2-ol in place of trimethylsilylacetylene. Early work with the 2-methyl-3-butyn-2-ol route resulted in incomplete cleavage of the 2-hydroxypropyl group (acetone as a by-product) from an intermediate using sodium hydroxide. Recent work using a catalytic amount of sodium hydride resulted in near-quantitative cleavage of the 2-hydroxypropyl group (9).

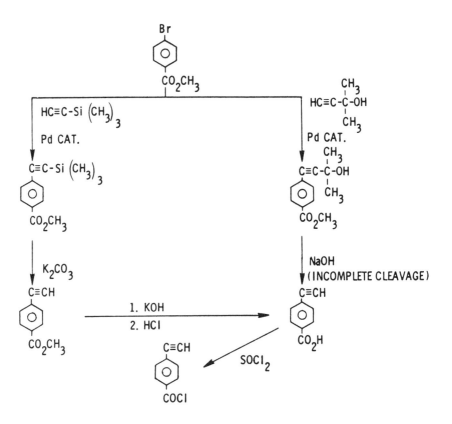

Figure 5

SYNTHESIS OF ETHYNYL-TERMINATED SULFONE (ETS)

An ethynyl-terminated sulfone (ETS) was prepared as shown in Figure 6 from the reaction of a hydroxy-terminated sulfone oligomer of $\bar{M}_n$ of ~12000 g/mole with a slight stoichiometric excess of 4-ethynylbenzoyl chloride as previously reported (5). Characterization of the ETS, the cured resin therefrom and UDEL® are presented in table 2. The 250°C cured ETS exhibited a higher Tg and better resistance to chloroform than UDEL®. This was due to the thermally induced reaction of the ethynyl group which caused some crosslinking. The swelling of the cured ETS in chloroform was expected since the linear sulfone portion of the molecule is relatively long and prone to attack by chloroform, an extremely aggressive solvent for polysulfones.

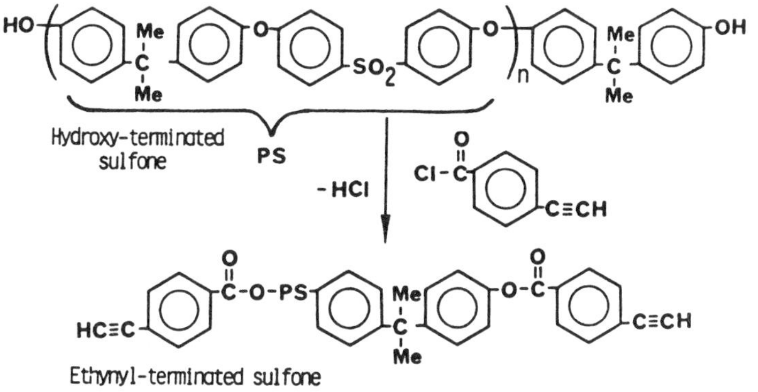

Figure 6

TABLE 2

Polymer	$\bar{M}_n$, g/mole	n_{inh}, dl/g	SEC Peak Retention Time, min	Tg, °C[1]	Solubility in Chloroform
ETS	~12,000	0.39	17.54	198	Soluble uncured, Swells when cured[1]
UDEL®	~25,000	0.44	17.14	192	Soluble

[1] After curing for 0.5 hr at 250°C in air

ISOTHERMAL AGING PERFORMANCE OF 250°C CURED FILMS

The thermooxidative stability of cured films (~0.002 in. thick) of UDEL® and two ETS of $\bar{M}_n$ = ~12,000 g/mole containing 20 and 80 ppm of palladium are shown in Figure 7. The effect of residual palladium which enhances the thermooxidative degradation is readily evident. The ETS with 20 and 80 ppm of palladium exhibited weight losses of 11.5 and 69.1 respectively after 90 hours at 250°C and 26.2 and 98.2 respectively after 64 hours at 288°C. The ETS exhibited relatively low weight losses at 177°C. UDEL® exhibited excellent stability at 177, 250 and 288°C. At 250°C, the UDEL® film melted to form a clump whereas the ETS films retained their form. Palladium was introduced during the preparation of 4-ethynylbenzoyl chloride. Recent work has involved the removal of the palladium from 4-ethynylbenzoyl chloride by the purification of one of the intermediates (9).

Aging Temp, °C (circulating air)	Time, hr	Weight Loss, %		
		UDEL®	ETS (20 ppm Pd)	ETS (80 ppm Pd)
177	1200	1.1	1.3	2.0
250	90	0.6	11.5	69.1
288	64	1.0	26.2	98.2

Figure 7

THIN-FILM PROPERTIES

The mechanical properties of thin films (~0.002 in. thick) of UDEL® and ETS prepared by solution casting onto plate glass and stage curing in air to a final temperature of 250°C and at 250°C for 0.5 hr are presented in Figure 8. Tensile strengths and moduli for the ETS film are higher than those of UDEL®, especially at 93°C. The UDEL® film was of better quality where elongation at 25° of 14% was obtained. The ETS film exhibited failure before it could elongate appreciably due to the poorer quality of the film. The UDEL® film was easily removed from the glass whereas severe problems were encountered in removing the ETS film from the glass. The ETS film could not be removed without tearing the glass, even upon soaking in water overnight. As a result, the ETS film had stress areas which served as points of weakness, causing premature failure, before the film could elongate appreciably. It was expected, however, that the elongation of the ETS film would be less than that of UDEL®.

Material	UDEL®		ETS ($\bar{M}_n$ = 12,000)	
Test temp., °C	25	93	25	93
Tensile strength, ksi	10.8 (10.6)[1]	8.2	12.1	9.6
Tensile modulus, ksi	320 (286)[1]	282	356	336
Elongation (break), %	14.1 (110)[1]	17.3	4.3	5.3

[1] Lit. (ref. 10) values for unoriented slot-cast thin film

Figure 8

PRELIMINARY UNIDIRECTIONAL CELION-6000 LAMINATE PROPERTIES

Unidirectional ETS and UDEL® prepregs were made by solution coating drum-wound Celion-6000 carbon/graphite fiber. The prepreg was vacuum dried at 100°C for 18 hours to reduce the volatile content to ~2%. Unidirectional laminates (3 in. x 7 in. x 0.060 and 0.110 in.) were fabricated in a stainless steel mold in a press starting at ambient temperature under 100 psi for UDEL® and 300 psi for ETS, heating under pressure to 316°C during ~316°C during ~45 minutes and maintaining at 316°C under pressure for 0.5 hour. Preliminary laminate properties are reported in Figure 9. The ETS laminate displayed slightly higher properties than UDEL® which may be attributed to better wetting of the filaments since the viscosity of the ETS solution used in prepreg preparation was lower than that of UDEL®. The flexural strengths of ETS at 177°C were higher than those of UDEL® although both systems exhibited thermoplastic failure at 177°C.

Test Conditions	Flexural st., psi x 10^3		Flexural mod., psi x 10^6		Short beam shear st., psi x 10^3	
	UDEL®	ETS	UDEL®	ETS	UDEL®	ETS
RT	189	197	18.6	19.4	8.5	9.3
93°C, 10 min.	177	192	17.1	18.9	8.1	8.9
177°C, 10 min.*	121	153	13.4	15.5	4.1	5.3
177°C, 1200 hr @ 177°C*	--	--	--	--	4.3	5.4

Resin Content: ~34%
*Thermoplastic failure

Figure 9

SOLVENT RESISTANCE (24-Hour SOAK)

The solvent resistances of various specimen forms of UDEL® and ETS after a 24-hour soak in different solvents are presented in Figure 10. UDEL® specimens exhibited sensitivity towards all the solvents whereas ETS was relatively unaffected by the solvents except for chloroform. Chloroform, like methylene chloride which is present in most paint strippers, is an aggressive solvent towards sulfone polymers. In chloroform, the area of the ETS film increased by ~55% due to swelling. Swelling is unacceptable for a composite matrix because this is obviously accompanied by severe strength loss.

Specimen Form	Solvent	Results UDEL®	ETS (Cured)
Stressed film	JP-4 Jet Fuel	Sl. Crazed	Unaffected
	Ethylene Glycol	Crazed	Unaffected
	Chevron Hyjet IV*	Badly Crazed	Unaffected
	Chloroform	Dissolved	Swelled
Hanging lap shear	Chevron Hyjet IV	~50% RT Strength Loss	~10% RT Strength Loss
	Chloroform	Dissolved	
Composite SBS	Chevron Hyjet IV	~20% RT Strength Loss	No RT Strength Loss
	Chloroform	Dissolved	Badly Swelled

*Hydraulic Fluid

Figure 10

FRACTURE ENERGY OF VARIOUS POLYMERS

The fracture energies of UDEL®, ETS, and various other polymers included for comparison are listed in Figure 11. Composite interlaminar fracture properties cannot be predicted based upon neat resin and fiber properties. High fracture energy of a neat resin does not necessarily mean a high interlaminar fracture energy in composite form. The ability of the resin to translate good composite properties must be determined. Several factors are important such as fiber/resin interface and residual stresses. As a result, it would be improper to predict the interlaminar fracture energy of an ETS composite based upon neat resin fracture energy. The cured ETS had a modest fracture energy of 1700 J/M^2, half the value of UDEL®.

POLYMER	FRACTURE ENERGY, J/M^2 [1]
UDEL® (UC Polysulfone)	3400^2
ETS ($\bar{M}_n$ = 12,000 g/mole)	1700^2
ULTEM® (GE Polyimide)	3700^2
121°C Cure Rubber Toughened Epoxy (Hexcel F-185)	5100^3
121°C Cure Epoxy (Hexcel 205)	270^3
177°C Cure Epoxy (Narmco 5208)	76^2

[1] Compact Tension Specimens.

[2] Courtesy of D. Hunston, National Bureau of Standards, private communication.

[3] Taken from reference 11.

Figure 11

SULFONE/ESTER POLYMER CONTAINING PENDENT ETHYNYL GROUPS

A hydroxy-terminated sulfone oligomer ($\bar{M}_n$ = ~4000 g/mole) was reacted with 5-(4-ethynylphenoxy)isophthaloyl chloride to yield a sulfone/ester polymer containing pendent ethynyl groups as shown in Figure 12. The physical properties of the sulfone/ester polymer and that of UDEL®, included for comparison, are presented in table 3. The cured sulfone/ester polymer exhibited better solvent resistance than UDEL® but not as good as anticipated. The crosslink density of the cured sulfone/ester polymer was not high enough to provide the degree of solvent resistance towards chloroform as observed for an ETS with a linear sulfone segment of ~3000 g/mole (5). The cured sulfone/ester polymer has a Tg higher than that of UDEL® presumably due to higher intermolecular association and not solely crosslinking.

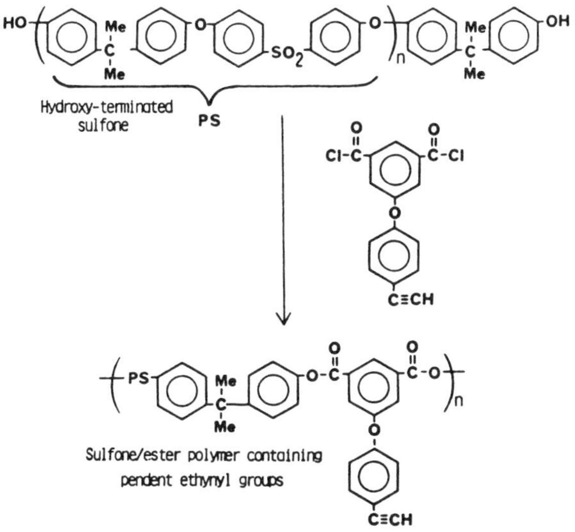

Figure 12

TABLE 3

Polymer	n_{inh}, dl/g[1]	SEC Peak Retention Time, min[1]	Tg of Cured Polymer, °C[1]	Chloroform Solubility of Cured Polymer Film[2]
Sulfone/Ester with Pendent Ethynyl Groups	0.34	17.55	204	Pronounced Swelling
UDEL®	0.44	17.14	192	Soluble

[1] See Figure 4.
[2] Cured 0.5 hour at 250°C in air.

FILM PROPERTIES OF SULFONE/ESTER POLYMER AND UDEL®

Preliminary mechanical properties of 250°C cured thin films (0.002 in. thick) of a sulfone/ester polymer containing pendent ethynyl groups and UDEL® are reported in Figure 13. The properties are essentially identical except that the sulfone/ester polymer showed no loss in tensile modulus at 93°C. The low elongation of the sulfone/ester polymer is not due solely to crosslinking but attributed to the poor quality of the film specimens. The films were cast on plate glass. The sulfone/ester polymer film tore the glass surface which caused stress areas in the film.

Polymer	UDEL®		Sulfone/Ester With Ethynyl Groups	
Test Temp, °C	25	93	25	93
Tensile Strength, ksi	10.8 (10.6)[1]	8.2	11.0	8.4
Tensile Modulus, ksi	320 (286)[1]	282	318.5	319.7
Elongation (Break), %	14.1 (110)[1]	17.3	4.6	3.3

[1]Taken from reference 10.

Figure 13

ETHYNYL-TERMINATED POLYARYLATES

Similar to the synthesis of ETS, ethynyl-terminated polyarylates (aromatic polyesters) were prepared from the reaction of hydroxy-terminated polyarylates of different $\bar{M}_n$ with 4-ethynylbenzoyl chloride (8) as shown in Figure 14. Characterization of the various ethynyl-terminated polyarylates and ARDEL® (a commercial aromatic polyester from Union Carbide Corporation) is presented in table 4. As the length of the linear ester segment increased in the cured polymers, the Tg decreased and the chloroform sensitivity increased due to lower crosslink density.

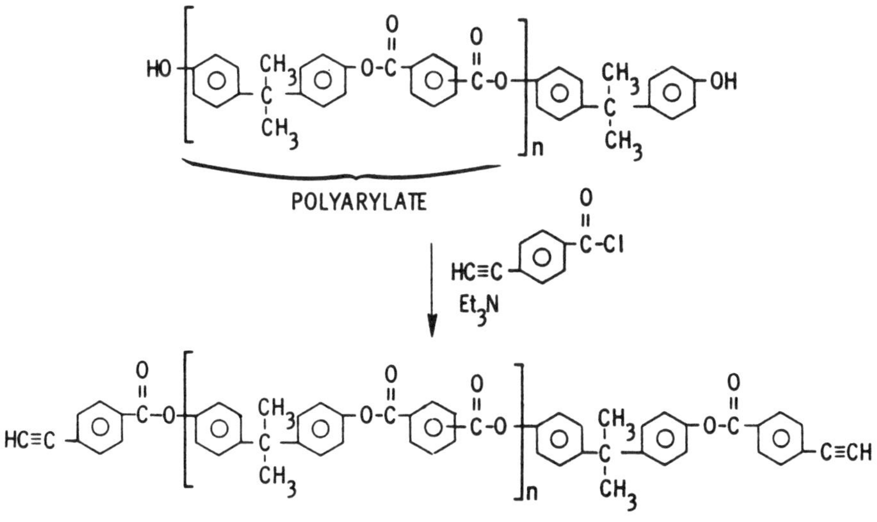

Figure 14

TABLE 4

$\bar{M}_n$, g/mole	η_{inh}, dl/g	GPC Peak Retention Time, min	Tg of Cured Polymer, °C	$CHCl_3$ Solubility of Cured Polymer
2,500	0.30	18.53	218	Very Sl. Swelling
5,000	0.34	18.23	215	Sl. Swelling
7,500	0.42	17.86	207	Swelling
10,000	0.47	17.45	203	Swelling
24,000 (ARDEL®)	0.59	17.21	197	Soluble

PHENOXY RESINS CONTAINING PENDENT ETHYNYL GROUPS

A commercial phenoxy resin (PKHH, Union Carbide Corp.) was systematically modified (12) by introducing different amounts of pendent ethynyl groups by reaction of the pendent hydroxy group on the phenoxy resin with 4-ethynylbenzoyl chloride and benzoyl chloride as shown in Figure 15. Characterization of the various compositions is presented in table 5. As the amount of pendent ethynyl groups in the phenoxy and accordingly crosslink density in the cured resin increased, the Tg and the resistance to chloroform also increased but the flexibility of films decreased.

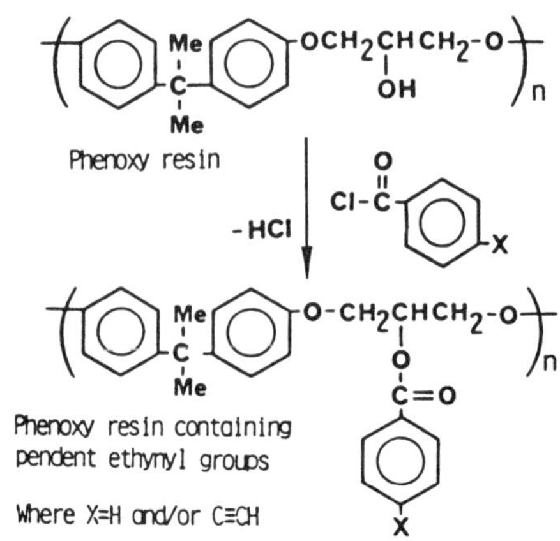

Figure 15

TABLE 5

Composition of Phenoxy		n_{inh}, dl/g	GPC Peak Retention Time, min	Tg of Cured Resin, °C	Film Flexibility	Chloroform Solubility of Cured Film (% increase in area)
Mole % benzoyloxy	Mole % ethynyl-benzoyloxy					
100	0	0.40	17.31	87	Very Flexible	Soluble
90	10	0.46	17.33	91	Very Flexible	>100% Swelling
66	34	0.48	17.25	110	Flexible	-50% Swelling
66	34	--	--	145	Brittle	<10% Swelling
34	66	0.47	17.30	150	Moderately Flexible	-30% Swelling
0	100	0.52	17.34	235	Brittle	Insoluble
PKHH	--	0.40	17.48	101	Very Flexible	Soluble

USE OF A COREACTANT TO INCREASE CROSSLINK DENSITY

The chemical structure and the DSC curve of a coreactant are shown in Figure 16. This coreactant was prepared from the reaction of bis-phenol A and 4-ethynylbenzoyl chloride. The DSC curve shows a melting endotherm peaking at 189°C and an exothermic peak at 219°C due to reaction of the ethynyl groups. Ten and thirty weight percents of the coreactant were blended with a phenoxy resin where 34% of the pendent hydroxyl groups had been converted to ethynylbenzoxyloxy group and the remaining 66% to benzoyloxy groups. The properties of the modified phenoxy resin without the coreactant and the blends are reported in table 6. As the amount of coreactant is increased, the cured blends showed an increase in the Tg, less flexibility in the film and better resistance to chloroform. The changes are attributed to higher crosslink density.

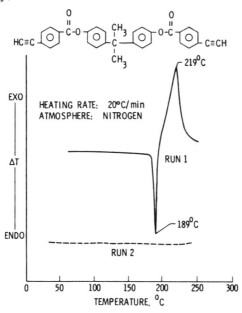

Figure 16

TABLE 6

Weight % of Coreactant	Tg of Cured Resin, °C	Film Flexibility	Chloroform Solubility of Cured Film (% Increase in Area)
0	110	Flexible	~ 50% Swelling
10	114	Moderately Flexible	~ 20% Swelling
30	145	Brittle	~ < 10% Swelling

CONCLUSIONS

o The following approaches improved the solvent resistance and raised the Tg of thermoplastics:

   o End-capping oligomers with ethynyl groups
   o Incorporating ethynyl groups pendent along the polymer chain
   o Coreacting polymers containing pendent ethynyl groups with a low molecular weight diethynyl compound

o Film and composite properties of an ethynyl-terminated sulfone were better than those of UDEL®

o Fracture energy of an ethynyl-terminated sulfone was lower than that of UDEL®

o Residual palladium in the cured ethynyl-terminated sulfone lowered the thermooxidative stability of the cured resin

o The properties of a phenoxy resin were altered considerably by placing pendent ethynyl groups along the polymer chain

o Property trade-offs must be considered when thermoplastics are modified via reactant groups

REFERENCES

1. G. A. Loughran, A. Wereta and F. E. Arnold, U. S. Pat. 4,108,926 (1978).

2. R. Vismanathan and J. E. McGrath, Polym. Prepr. 20 (2), 365 (1979).

3. C. H. Sheppard, E. E. House and M. Stander, Soc. Plast. Ind. Inc., 36th Ann. Conf. Prepr., Session 17-B, page 1 (1980).

4. C. H. Sheppard, E. E. House and M. Stander, Nat'l SAMPE Conf. Ser. 14, 70 (1982).

5. P. M. Hergenrother, J. Polym. Sci. Polym. Chem. Ed. 20, 3131 (1982).

6. J. E. McGrath, T. C. Ward, E. Shchori and A. J. Wnuk, Polym. Eng. Sci. 17, 647 (1977).

7. A. J. Wnuk, T. F. Davidson and J. E. McGrath, Polym. Prepr. 19 (1), 506 (1978).

8. S. J. Havens and P. M. Hergenrother, Polym. Prepr. 24 (2), 16 (1983).

9. S. J. Havens and P. M. Hergenrother, submitted to J. Org. Chem., August 1984.

10. R. N. Johnson in Encycl. Polym. Sci. and Tech. (H. F. Mark, N. G. Gaylord and N. M. Bikales, eds.) Vol. 11, Wiley-Interscience, New York, 1969, p. 447.

11. W. D. Bascom, J. L. Bitner and A. R. Siebert, Composites, 11, (9), 1980.

12. P. M. Hergenrother, B. J. Jensen, and S. J. Havens, Phenoxy Resins Containing Pendent Ethynl Groups, NASA TM-85747, January 1984.

THERMOPLASTIC/MELT-PROCESSABLE POLYIMIDES

T.L. St. Clair and H.D. Burks

NASA Langley Research Center
Hampton, VA

INTRODUCTION

Linear aromatic polyimides since their inception in the 1960's have been regarded as difficult materials to process into useable shapes. During the last 15 years many significant advances have been made in the processing of these materials. One particular advance that holds promise for the structural materials field has been the synthesis of polyimides that can be processed as thermoplastics. The initial materials that were processable in this manner required very high temperatures and pressures for proper consolidation. This paper describes some of the advancements that have been made at NASA Langley in developing aromatic polyimides that are facile to process. The primary objective has been to lower processing temperatures and/or pressures through the synthesis of polyimides with diluted imide content and/or with greater chain flexibility.

OBJECTIVE: TO LOWER THE PROCESSING TEMPERATURE OF LINEAR AROMATIC POLYIMIDES

APPROACH: A SYNTHESIS PROGRAM WHEREBY THE IMIDE CONTENT IS DILUTED & THE CHAIN FLEXIBILITY IS INCREASED

LARC TPI

A linear thermoplastic polyimide, LARC TPI, has been synthesized, characterized and developed for a variety of high-temperature applications. In its fully imidized form this material can be used as an adhesive for bonding metals such as titanium, aluminum, copper, brass, and stainless steel. LARC TPI is being evaluated as a thermoplastic adhesive for bonding titanium (6-aluminum-4-vanadium alloy) and to-date has shown no loss in properties after 25,000 hours of air-aging at 450°F. It is also being developed commercially as a laminating agent for bonding large pieces of polyimide film to metals to produce 100% void-free flexible circuits. This polymer has also been evaluated as a molding powder, composite matrix resin, high-temperature film, and fiber (ref. 1).

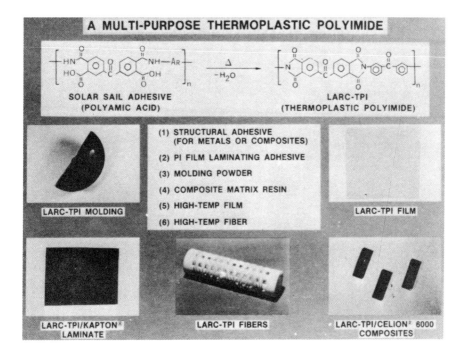

THERMOPLASTIC POLYIMIDESULFONE

Aromatic polysulfones, a class of high-temperature engineering thermoplastics, have a major deficiency in their tendency to swell and dissolve in many common solvents. This solvation can cause structural components which are fabricated from these polymers to be susceptible to damage by these solvents and thereby lose their structural integrity.

Aromatic polyimides, conversely, are a class of polymers which are known to be resistant to solvents, but they are generally not processable via thermoplastic means. These polyimides are known to be exceptionally thermally stable and like polysulfones and other thermoplastics their use temperature is governed by the softening temperature of each system.

A novel polymer system that possesses the processability of the polysulfones and the solvent resistance of the polyimides has been synthesized and characterized as a film, unfilled molding, and adhesive. The structure of this polyimidesulfone is shown below along with some adhesive and molding data (ref. 2).

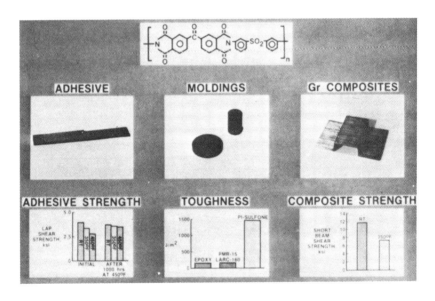

HOT-MELT PROCESSABLE POLYIMIDE

Linear aromatic polyimides are a class of polymers which are generally not processable via conventional thermoplastic or hot-melt techniques. This class of polymer is, however, exceptionally thermally stable and has high glass transition temperatures. It is also resistant to attack by common organic solvents.

Linear aromatic polyphenylene oxides and sulfides, on the other hand, are more easily processed than the polyimides, generally exhibit lower glass transition temperatures, and still have relatively good thermal stability, although not equal to the polyimides. These systems also do not possess solvent resistance equal to the polyimides.

A novel linear aromatic polyphenylene ethersulfideimide (PPX-PI) has been synthesized which has some of the favorable characteristics of each parent system. The polymer has been molded, used as a resin, and cast into thin films. A limited characterization indicates this system can be processed via conventional thermoplastic techniques and may have a wide variety of applications (ref. 3).

The synthesis and characterization of this novel system is the primary subject for the remainder of this paper. The overall characteristics of this polymer system make it very attractive for use as a composite matrix resin.

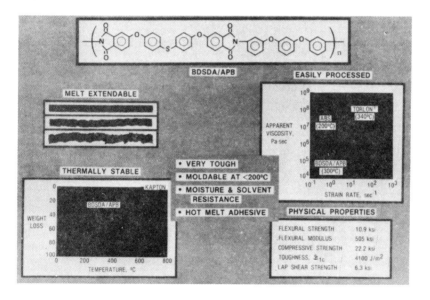

POLYMER SYNTHESIS SCHEME

The PPX-PI system has been prepared via the scheme shown below. The reaction of the monomers is carried out at room temperature in the solvent diglyme. The resulting polyamide-acid is soluble in this solvent, but when it is converted to the imide the polymer becomes insoluble. This phenomenum allows the polymer to be handled in solution form if this is desirable (prepregging) and after conversion to the imide form the polymer no longer has affinity for the solvent. This is in sharp contrast to other thermoplastics because they tend to tenaciously hold onto the solvent in which they were dissolved. The key point that must be emphasized is the subject PPX-PI system is one class of polymer in solution that becomes another class of polymer when the solvent is thermally removed. Both systems are thermoplastic.

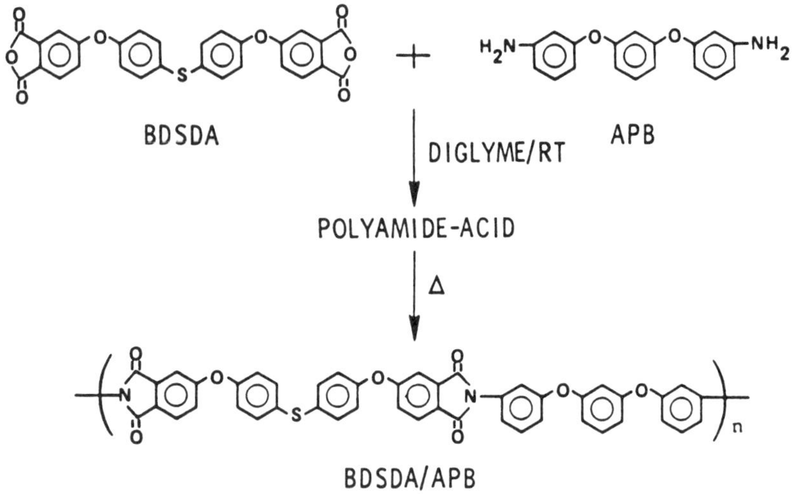

MECHANICAL PROPERTIES UNFILLED BDSDA/APB

The properties of the unfilled polymer are shown below. The flexural strength of 10.9 ksi is typical of thermoplastic polymers. The flexural modulus of this material is quite high when compared to conventional thermoplastics. In fact this modulus is more characteristic of thermoset systems. In contrast is the G_{Ic} value for this material. The value of 4100 J/m^2 is exceptionally high compared to these thermoset systems (< 100 J/m^2). Because of the combination of high modulus and high G_{Ic} (toughness indicator) this system should be attractive for the fabrication of impact-resistant structures.

FLEXURAL STRENGTH	75.1 MPa (10.9 ksi)
FLEXURAL MODULUS	3.48 GPa (505 ksi)
COMPRESSIVE STRENGTH	153 MPa (22.2 ksi)
CRITICAL RATE OF RELEASE OF STRAIN ENERGY, G_{Ic}	4100 J/m^2
LAP SHEAR STRENGTH (Ti/Ti)*	
LOW HEATING RATE (5°C/min)	40.3 MPa (5.85 ksi)
HIGH HEATING RATE (22°C/min)	43.4 MPa (6.30 ksi)

* BONDED AND TESTED ACCORDING TO ASTM STANDARD D 1002-72 (REF. 4)

THERMOOXIDATIVE STABILITY COMPARISON

The PPX-PI (BDSDA/APB) system has very good thermooxidative stability. When a 40-μm thick piece of this material was subjected to a dynamic thermogravimetric weight analysis at a standard 2.5°C/minute heating rate, the subject material had a weight loss profile similar to a 25-μm thick Kapton® standard that is used for instrument calibration. The approximate 10-20°C difference in decomposition temperature is significant and indicates that this material would not withstand elevated temperature (> 200°C) exposures as well as Kapton®.*

* Kapton is a registered trademark for a DuPont polyimide film.

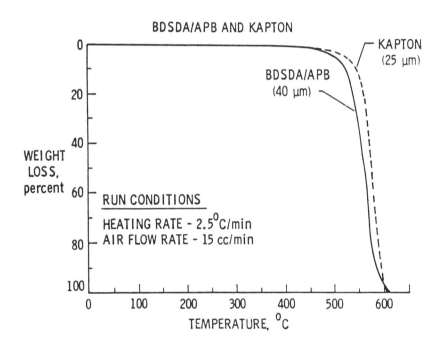

VISCOSITY AT MIDRANGE PROCESSING TEMPERATURE

This figure compares the change in apparent viscosity with strain rate at the midrange processing temperature for the subject polymer (BDSDA/APB), commercially available Torlon*, and a typical widely used ABS resin (ref. 5). This comparison is made because no data have been generated on a linear aromatic polyimide system prior to this BDSDA/APB study. At low strain rates the BDSDA/APB exhibits a considerably lower melt viscosity (i.e., lower processing pressure) than Torlon or ABS resin and maintains this relationship even at the higher strain rates. This would indicate it to be a somewhat more easily processable material.

*Torlon is the registered trademark of an Amoco poly(amide-imide).

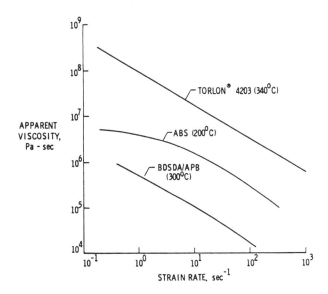

APPARENT VISCOSITY FOR PROCESSING PARAMETERS

The apparent viscosity as a function of strain rate at various temperatures is shown for the strain rates encountered in different industrial processes. The apparent viscosity was calculated by dividing the flow stress by the strain rate. As the strain rate was calculated from the volumetric flow data and was not corrected to obtain the wall rate, the viscosity is an apparent rather than a true viscosity (ref. 6). The BDSDA/APB polymer should be processable via compression molding and calendering techniques. However, no conclusions can be drawn concerning the extrudability of the polymer above a strain rate of 135 sec^{-1} due to the stress and strain rate limitations of our rheometer in its present configuration.

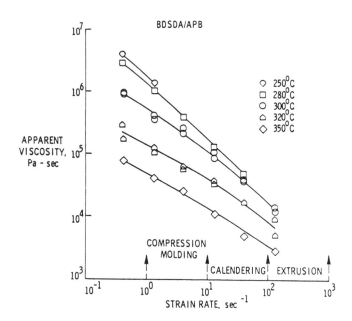

CHEMICAL RESISTANCE

The chemical resistance of BDSDA/APB thin film (40μm thick) to six common solvents was determined and the results listed in this figure. Methylethyl ketone, cyclohexanone, xylene, and tricresylphosphate had no visible effect on the film and there was no change in T_g (apparent). Methylene chloride and cresol caused severe swelling and T_g measurements were not possible, although the methylene chloride-soaked film did maintain sufficient integrity as a film to allow mounting in the T_g fixture.

BDSDA/APB FILM

SOLVENT	EFFECT	CHANGE IN T_g, (APPARENT), °C*
METHYLETHYL KETONE	NONE	NONE
CYCLOHEXANONE	NONE	NONE
XYLENE	NONE	NONE
TRICRESYLPHOSPHATE	NONE	NONE
METHYLENE CHLORIDE	SWELLED	NOT DETERMINED
CRESOL	SWELLED	NOT DETERMINED

40-μm THICK FILM
*THERMOMECHANICAL ANALYSIS OF SOLVENT-LADEN FILMS SOAKED 72 hrs AT ROOM TEMPERATURE AND BLOTTED DRY

ENDCAPPED PROCESSABLE POLYIMIDE

In an attempt to improve the processability of the BDSDA/APB polyimide system a series of endcapped analogues were prepared. These were accomplished by substituting the monofunctional phthalic anhydride for some of the BDSDA. The general synthesis scheme used was to decrease the difunctional BDSDA by one mole percent and add the phthalic anhydride at a two-mole percent level. A two-for-one substitution is used because the anhydride molar equivalence must be held equal to the amine molar equivalence. In subsequent figures the 1%, 2%, and 4% endcapped systems refer to the molar amount of phthalic anhydride (ref. 7).

BDSDA/APB

EFFECT OF MOLECULAR WEIGHT ON VISCOSITY

An objective of this research was to determine the effect of M_n on melt-flow properties of this polymer (BDSDA/APB). Based on the work of Fox and Flory (ref. 8) it was expected that as the M_n decreased the melt-flow would increase. However, their work was on another polymer and the degree of change must be determined for each individual polymer system. In the endcapped polyimide study there was a direct relationship between M_n and apparent viscosity as shown in the figure. At 250°C there was no deviation from linearity, but at the higher extrusion temperature of 280°C the viscosity dropped precipitiously for the systems with an M_n below 10,000. The apparent viscosity of the 4% endcapped material was too low at 280°C for the capillary rheometer to measure. Particularly noteworthy was the drop of two orders in magnitude for the apparent viscosity at 250°C when the M_n changed from approximately 14,000 to 8,700 amu. This information should be valuable in optimizing processabiity.

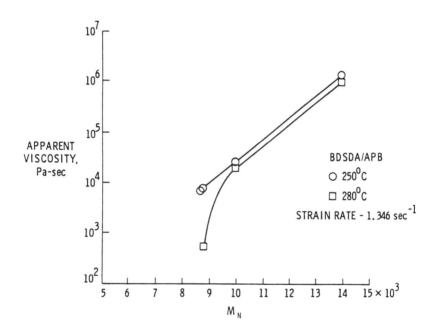

EFFECT OF MOLECULAR WEIGHT ON TOUGHNESS

When moldings were prepared from the different molecular weight polymers, it was obvious that the higher M_n systems resisted cracking more than the lower M_n systems. The G_{Ic} data proved that a considerable loss in fracture resistance does occur as the molecular weight decreases. This relationship is shown in the figure. Of particular interest was that over the range tested the G_{Ic} value was linearly related to the number-average molecular weight. The steep slope of this relationship was quite surprising. A change in M_n from 14,000 to 10,000 amu resulted in nearly a 3000 J/m² loss in G_{Ic}.

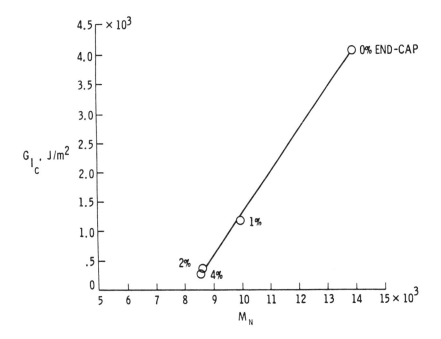

SYNTHESIS OF TWO PROCESSABLE POLYIMIDES

In an effort to further research the area two additional hot-melt processable polyimides have been prepared (ref. 9). These materials have oxygen, sulfur and bridges which link the aromatic rings through both the para and meta positions. The synthesis of the amide-acid polymer was accomplished at room temperature in diglyme as the solvent. These polymers were precipitated in water and after air-drying overnight they were imidized in a forced-air oven at 300°C.

VISCOSITY/STRAIN RATE DATA

The apparent viscosity as a function of strain rate is shown in the figure for the two new polyimides. The BDSDA/APB viscosity/strain rate characteristics are shown for comparison. The new materials exhibited continuous flow over the strain rate regime indicated. However, this flow is not continuous to the higher strain rate levels as it was for the BDSDA/APB. Both of the new materials required a higher processing temperature and exhibited higher viscosity values. Nevertheless, both materials show promise for hot-melt processing.

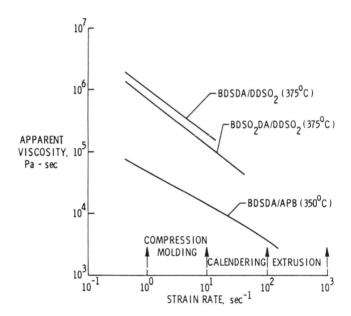

COMMERCIALLY PROMISING PROCESSABLE POLYIMIDES

Presently there are two polyimides that are available and attractive for hot-melt processing into fiber-reinforced laminates. These are Ultem® from General Electric and the NASA Langley BDSDA/ODA (LARC-ODA). The LARC-ODA is available in experimental quantities from M&T Chemicals (Rahway, NJ). This material has also been prepregged onto woven graphite by American Cyanamid (Havre de Grace, MD). Both systems exhibit very high G_{IC} values, 3700 J/m^2 for Ultem and 5400 J/m^2 for LARC-ODA.

Unfortunately, the Ultem is only available in the imidized form which causes difficulty in the preparation of composites for two reasons. First, when attempts have been made to hot melt impregnate fiber, poor wet-out of the fiber occurs. Second, any solvent used to dissolve the Ultem for solvent impregnation is tenaciously held by the polymer during the entire fabrication operation. Since the LARC-ODA is made as an amide acid it is soluble in solvents that are easily excluded when the material converts to the insoluble imide.

ULTEM — GENERAL ELECTRIC

BDSDA—ODA   LaRC

SUMMARY

Several polyimides have been prepared which show promise for aircraft composite applications. This has been achieved through a systematic polymer synthesis program where the glass transition temperatures have been greatly lowered when compared to the older polyimide systems. Several of the materials have been shown to be hot-melt processable and are attractive matrix resin candidates especially in light of their high G_{Ic} values. At least two of these polyimides are available for evaluation and others are on the research horizon.

o SEVERAL PROCESSABLE PIs HAVE BEEN PREPARED

o A SYSTEMATIC DECREASE IN PROCESSING TEMPERATURE HAS BEEN ACHIEVED

o PROMISING COMMERCIAL CANDIDATES HAVE BEEN IDENTIFIED

o SCALE-UP OF SEVERAL POLYMERS HAS BEEN ACCOMPLISHED

REFERENCES

1. St. Clair, A. K. and St. Clair, T. L.: A Multi-Purpose Thermoplastic Polyimide. Material and Process Applications: Land, Sea, Air, and Space, 26th National SAMPE Symposium and Exhibition 26, pp. 165-179, 1981.

2. St. Clair, T. L. and Yamaki, D. A.: A Thermoplastic Polyimidesulfone. NASA TM 84574, November 1982.

3. Burks, H. D. and St. Clair, T. L.: Synthesis and Characterization of a Melt Processable Polyimide. NASA TM 84494, May 1982.

4. Standard Test Method for Strength Properties of Adhesives in Shear by Tension Loading (Metal-to-Metal), ASTM Standard D 1002-72, 1982 Annual Book of ASTM Standards, Part 10, 1982.

5. Torlon Applications Guide, Amoco Chemical Corp., p. 11, Feb. 1979.

6. VanWazer, J. R.; Lyons, L. W.; Kim, K. Y.; and Colwell, R. E.: Viscosity and Flow Measurements, Interscience Publishers, New York, NY, p. 193, 1963.

7. Burks, H. D. and St. Clair, T. L.: The Effect of Molecular Weight on the Melt Viscosity and Fracture Energy of BDSDA/APB. J. of Applied Polymer Science, vol. 29, pp. 1027-1030, March 1984.

8. Fox, T. G. and Flory, P. J.: Second-Order Transition Temperatures and Related Properties of Polystyrene. J. of Applied Physics, 21, p. 581, 1950.

9. Burks, H. D. and St. Clair, T. L.: Synthesis and Characterization of a Polyethersulfoneimide. NASA TM 84621, April 1983.

ALIPHATIC-AROMATIC HETEROCYCLICS AS POTENTIAL THERMOPLASTICS FOR COMPOSITE MATRICES

Chad B. Delano and Charles J. Kiskiras

Acurex Corporation
Aerotherm Division
Mountain View, California

INTRODUCTION

The successful development of impact- and solvent-resistant thermoplastic systems for glass and graphite composites is particularly attractive because of the demonstrated streamlined manufacturability of such composites. Hypothetically, thermoplastics only require simple heating and cooling cycles for component manufacture, whereas thermosets require precise, and possibly extended heating schedules which must be consistent with the cure chemistry. The sensitivity of the majority of existing thermoplastics to aircraft fluids and other solvents preempts their serious consideration in aircraft components. This is the basic reason that Acurex proposed insolubility in common solvents as the starting point for this NASA Langley sponsored program. The target properties for the new thermoplastic resin are summarized in Figure 1.

- PREPREG PROPERTIES
    - -- USE OF CONVENTIONAL PREPREGGING EQUIPMENT
    - -- 6+ MONTHS SHELF STABILITY
    - -- PROCESSABILITY: 316ºC, 0.69 MPa (600ºC, 100 PSI)
    - -- THERMOFORMABILITY

- COMPOSITE PROPERTIES
    - -- -54ºC TO 93ºC CAPABILITY
    - -- GOOD MECHANICAL PROPERTIES
    - -- GOOD ENVIRONMENTAL PROPERTIES (UNDER STRESS)
    - -- IMPACT RESISTANCE

---

*IN PROGRESS NAS1-16808 -- DEVELOPMENT OF AN IMPACT AND SOLVENT RESISTANT COMPOSITE MATRIX

Figure 1

PBI-8

Several aliphatic-aromatic heterocyclic polymers have been reported in the literature; however, systematic studies of their properties as thermoplastics are not readily available. Acurex was attracted to this class of polymers for the excellent solvent resistance demonstrated by the polybenzimidazole from 3,3',4,4'-tetraaminobiphenyl with sebacic acid (PBI-8). Its solvent resistance is compared to the solvent resistance of polysulfone and polyphenylene oxide (PPO) in Figure 2.

Property	PBI-8[a]	Polysulfone	PPO
Heat deflection temperature, °F	420	345	375
Thermal coefficient of expansion in./in./°F	4.1×10^{-5}	3.1×10^{-5}	2.9×10^{-5}
Specific gravity	1.17	1.24	1.06
Water absorption, percent (24 hr)	0.28	0.22	0.06
Tensile strength, psi at 72°F	11,000	10,200	11,000
at 275°F	4,500		
Tensile elongation, percent at 72°F	5	50 to 100	50 to 80
Modulus, psi at 72°C	3.4×10^5	3.9×10^5	3.7×10^5
Solvent resistance[b]			
Aliphatic hydrocarbons	N	N	N
Aromatic hydrocarbons	N	P	S
Chlorinated hydrocarbons	N	S	S
Ketones	N	S	P

[a] PBI-8 = Acurex's melt processible PBI
[b] N = no effect; P = partially soluble; S = soluble

Figure 2

APPROACH

The Acurex approach included selection of several flexible and rigid segments for two classes of polymers as outlined in Figure 3.

- SOFT (FLEXIBLE) SEGMENT-RIGID SEGMENT POLYMERS
- ALIPHATIC-AROMATIC HETEROCYCLIC POLYMERS
    -- ALIPHATIC POLYIMIDES
    -- ALIPHATIC POLYBENZIMIDAZOLES

Figure 3

POLYIMIDE STRUCTURES

The flexible and rigid segments selected for investigation of polyimides are indicated in Figure 4. The dianhydrides investigated were pyromellitic dianhydride (PMDA), 3,3'4,4'-benzophenonetetracarboxylic dianhydride (BTDA) and 5-(2,5-diketotetrahydrofuryl)-3-methyl-3-cyclohexene-1,2-dicarboxylic dianhydride (MCTC). The diamines investigated were 1,6-hexanediamine, 1,8-octanediamine (ODA) and 1,12-dodecanediamine.

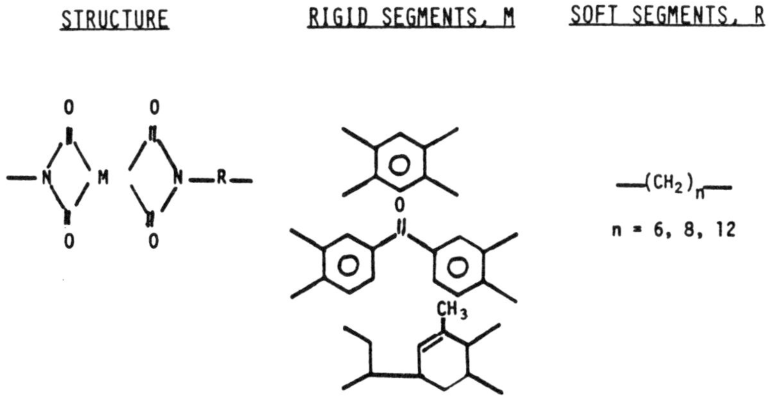

Figure 4

LINEAR ALIPHATIC POLYBENZIMIDAZOLES

We were successful in melt polymerizing several polybenzimidazoles which were subsequently characterized for both Tg and solvent resistance. Inherent viscosities of the polymers from the two types of polybenzimidazoles are indicated in Figure 7.

REACTANTS	VISCOSITY, DL/GM
3,3',4,4'-TETRAAMINOBIPHENYL	
SUBERIC ACID	1.59 (H_2SO_4 -- 33% INSOLUBLES)
SEBACIC ACID	2.69 " 22% "
1,12-DODECANEDICARBOXYLIC ACID	2.31 " 7.2% "
4,4'-BIS(O-AMINOANILINO) BIPHENYL	
SUBERIC ACID	1.10 (H_2SO_4)
SEBACIC ACID	0.88 (CRESOL)
1,12-DODECANEDICARBOXYLIC ACID	0.68 (H_2SO_4 -- 3.5% INSOLUBLES)
METHYLHEXAHYDROPHTHALIC ANHYDRIDE	CLEAR 1.10; OPAQUE 1.62 (CRESOL)
PHTHALIC ANHYDRIDE	0.92 (H_2SO_4)

- MELT CONDENSATION PRODUCES TOUGH POLYMERS

Figure 7

FLOW BY TMA

We found that the flow properties of powdered resin samples could be determined conveniently in the laboratory by placing the powder between two DSC cups* as sketched in Figure 8 and generating the familiar TMA† curves. Initial trial molding temperatures for each polymer were selected from such curves and after obtaining molded resin specimens TMA curves were conducted at higher penetration pressures for heat distortion temperature simulation to provide indication of upper use temperatures for each polymer composition. This latter test procedure is also sketched in Figure 8.

---

*Small thin metal dishes used in differential scanning calorimetry (DSC)
†Thermomechanical analysis (TMA)

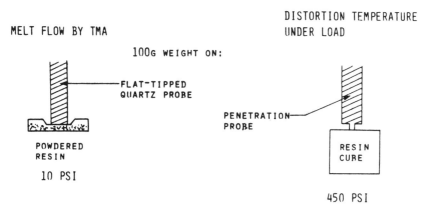

- PROCEDURES EMPLOYED FOR DETERMINING MOLDING CONDITIONS AND UPPER USE TEMPERATURES

Figure 8

FLOW CURVES FOR ALIPHATIC POLYIMIDES

Low pressure and distortion temperature under load (DTUL) curves for the PMDA-based polyimides are indicated in Figure 9. DSC analysis showed these polyimides to be crystalline. The T_m^* is indicated on the diagram by the triangle and is seen to control the flow properties of the polymer.

$^*T_m$ = crystalline melting temperature

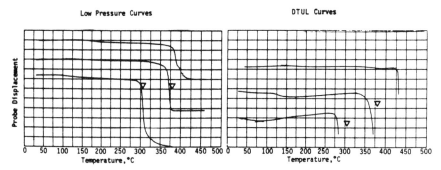

PMDA with C_6, C_8, and C_{12} diamines (top to bottom curves respectively)

- INVERSE RELATIONSHIP BETWEEN T_M AND DIAMINE LENGTH EXISTS
- 450 PSI COMPRESSIVE LOAD DID NOT PENETRATE UNTIL T_M REACHED

Figure 9

DSC CURVES FOR PMDA/1,12-DODECANEDIAMINE POLYIMIDE

The heating curve shown in Figure 10 shows an endotherm close to 315°C and an exotherm upon cooldown close to 230°C for this polyimide. Its crystallinity was confirmed by X-ray analysis. As indicated in the previous figure the degree of crystallinity is high enough to control its flow properties and support the 450 psi compressive load (DTUL measurement) to very high temperatures.

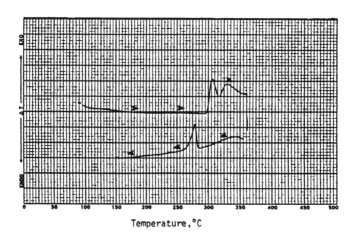

- **THERMAL CRYSTALLIZATION**

Figure 10

FLOW CURVES FOR ALIPHATIC POLYIMIDES

Low pressure flow and DTUL curves for the BTDA- and MCTC-based polyimides are provided in Figure 11. DSC analysis of these polymers provided Tg which is indicated on the diagrams. The Tg of these noncrystalline systems clearly controls their flow properties.

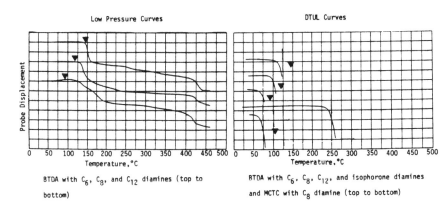

BTDA with C_6, C_8, and C_{12} diamines (top to bottom)

BTDA with C_6, C_8, C_{12}, and isophorone diamines and MCTC with C_8 diamine (top to bottom)

- BTDA POLYMERS DID NOT SHOW CRYSTALLINITY
- EXPECTED HIGH T_G OBTAINED WITH CYCLOALIPHATIC COMPARED TO LINEAR ALIPHATIC DIAMINE

Figure 11

FLOW CURVES FOR ALIPHATIC POLYBENZIMIDAZOLE

Low pressure flow and DTUL curves for the aliphatic polybenzimidazoles are shown in Figure 12. Unlike the flow curves for the polyimides shown in Figures 9 and 11 where low pressure and 450 psi pressure produce approximately the same flow properties, significant differences exist between the flow properties of the aliphatic polybenzimidazole under the two conditions. Lower molecular weight polymers frequently do flow under low pressure testing at lower temperatures. Thermal history of the polymers prior to flow testing appears to play an important role in the flow properties of these systems.

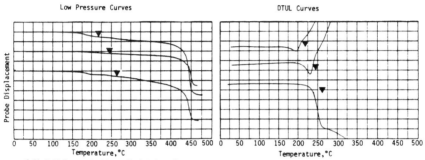

3,3',4,4'-Tetraaminobiphenyl with C_{12}, C_8, and C_6 dicarboxylic acids (to to bottom curves respectively)

- HIGH MOLECULAR WEIGHT POLYMERS DO NOT FLOW AT LOW PRESSURES UNTIL FAR ABOVE T_G
- 450 PSI COMPRESSIVE LOAD PENETRATED POLYMERS AT MUCH LOWER TEMPERATURES

Figure 12

FLOW CURVES FOR N-ARYLENEPOLYBENZIMIDAZOLES

Figure 13 provide the low pressure and DTUL curves of the aliphatic N-arylenepolybenzimidazoles. The Tg indicated on the diagrams appears to control the flow properties of these polymers except for the low pressure curve for the polymer from suberic acid which as prepared was crystalline. The molded polymer was not crystalline.

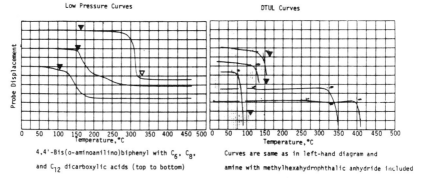

4,4'-Bis(o-aminoanilino)biphenyl with C_6, C_8, and C_{12} dicarboxylic acids (top to bottom)

Curves are same as in left-hand diagram and amine with methylhexahydrophthalic anhydride included

- INVERSE RELATIONSHIP BETWEEN FLOW TEMPERATURE AND DIAMINE LENGTH EXISTS
- HIGH FLOW TEMPERATURE PROPERTY OF SUBERIC ACID IS DUE TO CRYSTALLINITY
- EXPECTED HIGH T_G OBTAINED WITH CYCLOALIPHATIC COMPARED TO LINEAR ALIPHATIC DIACIDS

Figure 13

DRY AND WET POLYMER PROPERTIES OF POLYIMIDES

Twenty-four hour water boil properties of epoxy resins show weight gains of approximately 5 percent. The weight gains of most of the polyimide moldings are much lower than epoxy resins and the environmental moisture durability of these systems would be predicted to be superior to epoxy resins. These values are indicated in Figure 14. One usually associates high barcol hardness values with composite matrix resins and most of the aliphatic polyimides investigated do not give high hardness values.

POLYMER	SPECIFIC GRAVITY (g/cc)	HARDNESS OF MOLDING		DTUL MEASUREMENTS (°C)			INTERCEPT SHIFT DUE TO MOISTURE (°C)	SAMPLE WEIGHT GAIN DUE TO 24-HR WATER BOIL (PERCENT)
		BARCOL	SHORE D	DEVIATION FROM LINEARITY	INITIAL DRY INTERCEPT	24-HR WATER BOILED INTERCEPT		
POLYIMIDES								
PMDA, C_6 DIAMINE	1.38	a	a	148	428	278	-150	0.9
PMDA, C_8 DIAMINE	1.28	6-10	--	108	353	350	-3	0.2
PMDA, C_{12} DIAMINE	1.20	0	65	NOT CLEAR	278	282	+4	0.0
BTDA, C_6 DIAMINE	1.32	20	--	98	113	102	-11	3.9
BTDA, C_8 DIAMINE	1.28	0	65	79	98	85	-13	1.6
BTDA, C_{12} DIAMINE	1.21	0	45	59	72	63	-19	0.9
BTDA, ISOPHORONEDIAMINE	1.14	a	a	222	242	130	-112	3.6
BTDA, C_8 DIAMINE AND M-PHENYLENEDIAMINE		40	--	155	187	180	-7	0.9
NCTC, C_8 DIAMINE	1.12	0	60	48	67	62	-5	9.2

aSAMPLE BREAKS DURING HARDNESS TESTING

MOISTURE WEIGHT GAIN OF PMDA POLYMERS IS REMARKABLY LOW
MARKED IMPROVEMENT IN BARCOL HARDNESS OCCURS WITH M-PHENYLENEDIAMINE COREACTANT

Figure 14

DRY AND WET POLYMER PROPERTIES OF POLYBENZIMIDAZOLES

Twenty-four hour water boil of the aliphatic polybenzimidazoles does not appear to offer much improvement over incumbent epoxy resins except with the 1,12-dodecanedicarboxylic acid system. The N-arylenepolyimidazoles appear to offer improved moisture resistance over the classical polybenzimidazoles (Figure 15).

POLYMER	SPECIFIC GRAVITY (g/cc)	HARDNESS OF MOLDING		DTUL MEASUREMENTS (°C)			INTERCEPT SHIFT DUE TO MOISTURE (°C)	SAMPLE WEIGHT GAIN DUE TO 24-HR WATER BOIL (PERCENT)
				INITIAL DRY		24-HR WATER BOILED		
		BARCOL	SHORE D	DEVIATION FROM LINEARITY	INTERCEPT	INTERCEPT		
PBIs								
FROM C_6 DICARBOXYLIC ACID	1.14	45	--	197	235	126	-109	7.9
FROM C_8 DICARBOXYLIC ACID	1.12	30	--	185	207	127	-80	5.6
FROM C_{12} DICARBOXYLIC ACID	1.08	18	--	160	175	134	-41	4.3
N-ARYLENE PBIs								
FROM C_6 DICARBOXYLIC ACID	1.17	15	--	115	143	118	-25	3.1
FROM C_8 DICARBOXYLIC ACID	1.13	10	--	108	125	102	-23	3.6
FROM C_{12} DICARBOXYLIC ACID	1.12	a	75	58	75	63	-12	2.9
FROM PHTHALIC ANHYDRIDE	--	a	a	300	321	319	-2	1.2
FROM METHYLHEXAHYDROPHTHALIC ANHYDRIDE	1.10	a	a	260	400	407	+7	1.1

aSAMPLE BREAKS DURING HARDNESS TESTING

ALIPHATIC POLYBENZIMIDAZOLES HAVE GOOD BARCOL HARDNESSES

T_G LOSS DUE TO WATER BOIL IS NOT AS GOOD AS THE ALIPHATIC POLYIMIDES

Figure 15

SOLVENT SCREENING OF MOLDED SPECIMENS

As indicated in Figure 16 the unstressed solvent resistance of most of the polyimides is excellent when acetone or tricresylphosphate is the test solvent. Upon exposure to chloroform, however, only three polymers show good resistance to this solvent. Chloroform is considered to be very aggressive in comparison to other organic solvents in its ability to soften plastics. It is related to chlorinated paint strippers which must not degrade the composite.

POLYMER	SOLVENT, EXPOSURE TIME, PERCENT WEIGHT GAIN								
	ACETONE			CHLOROFORM			TCP		
	1 DAY	7 DAYS	50 DAYS	1 DAY	7 DAYS	50 DAYS	1 DAY	7 DAYS	50 DAYS
POLYIMIDES									
PMDA, C_6 DIAMINE	0	0	0	0	0	--	1	2	0
PMDA, C_8 DIAMINE	0	0	a	3	5	a	1	0	a
PMDA, C_{12} DIAMINE	1	1	3	12	30	29	0	0	0
BTDA, C_6 DIAMINE	0	0	2	SWELLS			1	0	0
BTDA, C_8 DIAMINE	2	6	11	SWELLS			0	0	0
BTDA, C_{12} DIAMINE	7	9	9	SWELLS			1	1	0
BTDA, ISOPHORONEDIAMINE	0	0	6	SWELLS			a	1	0
BTDA, C_8 DIAMINE AND M-PHENYLENE DIAMINE	1	--	1	2	--	6	0	--	0
MCTC, C_8 DIAMINE	SWELLS			DISSOLVED			SWELLS	DISSOLVED	

aSAMPLE USED FOR DTUL MEASUREMENT

COMMON SOLVENT RESISTANCE OF PMDA/ODA AND BTDA/ODA-MPDA POLYIMIDES IS EXCELLENT

Figure 16

SOLVENT SCREENING OF MOLDED POLYBENZIMIDAZOLES

As indicated in Figure 17 the two classes of aliphatic polybenzimidazoles showed completely different responses to chloroform exposure. The unmodified N-arylenepolybenzimidazoles do not appear as a class to be suitable for use in composites where exposure to paint strippers can occur compared to the classical polybenzimidazoles which are excellent in this regard.

POLYMER	SOLVENT, EXPOSURE TIME, PERCENT WEIGHT GAIN								
	ACETONE			CHLOROFORM			TCP		
	1 DAY	7 DAYS	50 DAYS	1 DAY	7 DAYS	50 DAYS	1 DAY	7 DAYS	50 DAYS
PBIs									
FROM C_6 DICARBOXYLIC ACID	0	0	0	0	0	--	4	4	5
FROM C_8 DICARBOXYLIC ACID	0	0	a	0(0)	0(0)	a	1(0)	2(0)	a(0)
FROM C_{12} DICARBOXYLIC ACID	0	0	a	2	4	a	1	1	a
N-ARYLENE PBIs									
FROM C_6 DICARBOXYLIC ACID	4	14	14	SWELLS			0	0	0
FROM C_8 DICARBOXYLIC ACID	11	18	16	DISSOLVED			-2	-1	0
FROM C_{12} DICARBOXYLIC ACID	23	14	-2	DISSOLVED			-2	-5	HALF DISSOLVED
FROM PHTHALIC ANHYDRIDE	6	--	--	SWELLS			--	--	--
FROM METHYLHEXAHYDROPHTHALIC ANHYDRIDE	15	15	2	DISSOLVED			3	4	4

a SAMPLE USED FOR DTUL MEASUREMENT

COMMON SOLVENT RESISTANCE OF PBI'S IS EXCELLENT
COMMON SOLVENT RESISTANCE OF N-ARYLENE PBI'S IS POOR

Figure 17

STRESSED SOLVENT RESISTANCE OF SELECTED POLYMERS

Three aliphatic-aromatic heterocyclic thermoplastic composites were selected for stressed solvent resistance testing. The composition of these polymers and the test method are indicated in Figure 18.

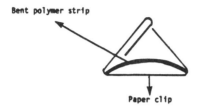

500 HR EXPOSURE TO ACETONE, CHLOROFORM AND TCP GAVE NO EVIDENCE OF FAILURE

-- POLYIMIDES FROM PMDA/1,8-OCTANE DIAMINE AND BTDA/1,8-OCTANE DIAMINE-M-PHENYLENE DIAMINE

-- POLYBENZIMIDAZOLE FROM 3,3',4,4'-TETRAAMINOBIPHENYL/SEBACIC ACID

Figure 18

POLYMER SELECTIONS

The properties of the three selected polymers which showed no effect under stressed solvent testing are summarized in Figure 19. Tensile properties of the systems are also indicated. The low tensile strength of the polyimide from PMDA with 1,8-octanediamine needs comment. The first molding of this system was very tough and nearly clear. The second molding was opaque and fracture sensitive. The latter molding was used in the tensile testing and gave premature failure. The modulus of this system needs to be redetermined. The polyimide from BTDA with 1,8-octanediamine and m-phenylene diamine was selected for further investigation.

	MOLD TEMP., °C	BARCOL HARDNESS	DRY[a] DTUL, °C	WET[a] DTUL, °C	TENSILE PROPERTIES
POLYIMIDES					STRENGTH, PSI/ELONGATION, %/MODULUS, PSI[b]
PMDA WITH C_8 DIAMINE	725	10	108 (353)	95 (350)	4,400/3/166,000
BTDA WITH C_8 DIAMINE AND M-PHENYLENE DIAMINE	510	40	155 (187)	148 (180)	17,000/14/177,000
POLYBENZIMIDAZOLES					
TAB WITH SEBACIC ACID	725	25	185 (207)	120 (126)	14,700/10/166,000

[a]DEVIATION FROM LINEARITY VALUES; INTERCEPT VALUES ARE IN PARENTHESIS
[b]TANGENT MODULUS, NOT INITIAL MODULUS

Figure 19

FIRST YEAR'S KEY PROGRAM RESULTS

The results of the first year's investigation on the aliphatic-aromatic heterocycle thermoplastics are summarized in Figure 20. The program results suggest that a number of these types of systems could be devised; for example, a benzoxazole rigid segment might be a candidate.

- DEVELOPED POLYMERIZATION FOR POLYIMIDES WHICH PRODUCES VERY HIGH MOLECULAR WEIGHT POLYMERS

- DEVELOPED POLYMERS WHICH OBTAIN COMMON SOLVENT INSENSITIVITY BY THREE MECHANISMS
  1. CRYSTALLINE POLYMERS
  2. INSOLUBLE SEGMENT COPOLYMERS
  3. HYDROGEN BONDED POLYMERS

- MILESTONE OF MOLDABLE, COMMON SOLVENT INSENSITIVITY ACHIEVED

- PROGRAM RESULTS SUGGEST THAT MANY SOLVENT RESISTANT, HIGH TENSILE ELONGATION RESINS CAN BE DEVELOPED

Figure 20

WORKING GROUP SUMMARY: MATRIX SYNTHESIS AND CHARACTERIZATION

P.M. Hergenrother, Chairman

The discussion in the synthesis section of the tough composite matrix workshop was extremely candid and open, with the majority of the 35 or so attendees actively participating. A list of questions had been compiled prior to the meeting to aid in the discussion. These questions, however, initiated a particular criticism of the workshop, namely, that no detailed resources or roadmap were presented and no objective was defined. In essence, the audience interested in synthesis wanted more specifics on where the NASA effort is headed, how NASA hopes to get there, and, when they arrive, what NASA is going to do with the technology.

Early in the discussion, the NASA position on exclusive licensing was explained. A detailed plan must be submitted which indicates how the invention will be developed to commercial application. Nonexclusive licenses are also available.

The following ten questions were addressed. The answers represent a consensus of the workshop participants.

1. Did the workshop provide an adequate overall review of the synthetic work? No, because the audience was not informed of the details of the overall effort, such as objective, resources, and how the program was tied together.

2. Is NASA addressing the proper subjects in its synthetic work? Yes, even though this response may appear to contradict the answer to the first question. The synthetic effort involves thermosets, including rubber and thermoplastic toughened systems, and thermoplastics, encompassing linear, lightly crosslinked, and semicrystalline types. It was generally expressed that the NASA synthetic effort serves a very useful purpose to industry and should be continued. Some minor changes will be forthcoming as a result of recommendations from the audience. These are discussed in answer to some of the following questions.

3. Is the synthetic effort too broad? The audience felt that NASA should continue to maintain a broad synthetic effort, performing scouting expeditions in an attempt to uncover new concepts and technology. It was emphasized that NASA is not developing products but is directing the effort toward developing fundamental information that can be used by industry, if desired, to develop products.

4. Should the synthetic effort concentrate on one area more than another? A broad synthetic effort should be maintained, but it was recommended that work on epoxies be de-emphasized. It was generally thought that epoxy work was adequately covered by industry. Another recommendation was that more fundamental work on crystalline polymers be done by NASA.

5. What will be the biggest advance in the synthesis of composite materials over the next 10 years? The answer was computer modeling and fiber interface improvement. Advances in computer modeling will assist the synthetic chemist in the design and preparation of a toughened composite matrix with the best overall combination of properties. Improvement at the fiber interface, primarily through new surface treatments, will help maximize the mechanical properties of the composite system.

6. Concerning thermoplastics, what is the maximum acceptable processing temperature and pressure? The answer was that any conditions that would be cost effective in reproducibly fabricating quality composites could be used. No restrictions were placed upon either temperature or pressure. However, the temperature must obviously be controlled to avoid thermal degradation of the thermoplastic.

7. Is resistance to paint strippers an absolute requirement of composite matrices? Yes; although there are other means of removing paint, paint stripping solvents will continue to find widespread use. Composite matrices under stress are prone to attack by aggressive liquids such as paint strippers. Even epoxies will swell under these conditions. Semicrystalline materials such as PEEK (crystalline polyetheretherketone) were reported to be unaffected by paint strippers.

8. Is crystallinity in a polymer an acceptable approach to a solvent-resistant tough composite matrix? The answer was a resounding yes, if the crystallinity can be controlled. This question was proposed because of conflicting views on crystalline polymers. Some researchers feel that the same degree and type of crystallinity cannot be uniformly obtained in a composite structure which varies in thickness because heat transfer and controlled cool-down rate are problems. Others feel that the degree of crystallinity in a matrix will change as a function of environmental exposure, particularly when under stress to hydraulic fluid at elevated temperatures. If the degree of crystallinity changes, the physical and mechanical properties can change accordingly. Although these are only a few of the unknowns of crystalline materials, it was felt that these questions would be answered in the future and also that crystalline polymers offer an extremely good potential as a tough solvent-resistant composite matrix.

9. Is there a need to predict neat resin properties from base polymer molecular structure? Yes, if neat resin properties can be correlated with composite matrix properties.

10. What evaluation techniques are needed to guide the synthetic effort? Standard tests are needed using small amounts of material to evaluate toughness and solvent and moisture resistance under stress, as well as other physical and mechanical properties. It was recommended that the synthetic chemist have a series of standard screening tests that can be used to help ascertain the potential of a new polymer. If a new polymer exhibits good performance in these initial screening tests, then costly scale-up work can be conducted to obtain larger quantities for more comprehensive evaluation. Unfortunately, as implied in question 9, currently no real correlations exist between neat resin properties and composite properties. Work at NASA as well as elsewhere is under way in an attempt to develop pertinent tests than can be used to determine if such a correlation is possible. Although it was agreed that standard screening tests are needed, no action was recommended to develop a series of such tests.

Part IV
Selected Additional Subjects

The information in Part IV is from *Advanced Materials Technology,* compiled by Charles P. Blankenship and Louis A. Teichman, NASA Langley Research Center, Hampton VA, 1982. The report is the proceedings of a seminar, sponsored by NASA Langley Research Center and the American Institute of Aeronautics and Astronautics, November 1982.

POLYMER MATRIX COMPOSITES RESEARCH AT NASA LEWIS RESEARCH CENTER

T.T. Serafini

NASA Lewis Research Center
Cleveland, OH

CURRENT PROGRAM THRUSTS

The objective of the polymer matrix composites research at the NASA Lewis Research Center is to develop technology for new generations of polymer matrix composites intended for application in advanced aeropropulsion systems. Other applications for the newly developed technology are in airframe and space structures. Research is performed in the following areas: 1) monomer/polymer synthesis, 2) polymer/composites characterization, 3) cure/degradation mechanisms, 4) polymer/composites processing, 5) environmental effects, and 6) thermo-mechanical properties. Emphasis is given not only to developing improved materials, but also to achieving a fundamental understanding of materials' behavior at the molecular level.

The current thrusts of the Lewis polymer matrix composites program are listed in figure 1. In keeping with the Lewis role as the lead center for propulsion research, the major emphasis of the Lewis program is in the area of engine applications. Research is also being conducted to develop matrix resins with improved toughness to support the inter-center program, with the lead center responsibility at the Langley Research Center, to develop improved composites for airframe applications. Highlights of recent progress in each of the program thrusts are discussed with the exception of the thrust to develop a 700° F matrix resin. Emphasis is given to reviewing key advances in improving the processability of PMR polyimides and the application of PMR-15 composites in engine static structures.

ENGINE APPLICATIONS

- DEVELOP MATRIX RESINS FOR USE AT 700° F
- DEVELOP PMR POLYIMIDES WITH IMPROVED PROCESSABILITY
- DEVELOP COMPOSITES MECHANICS METHODOLOGY TO PREDICT LIFE/DURABILITY OF COMPOSITES IN ENGINE ENVIRONMENTS
- ESTABLISH FABRICATION TECHNOLOGY FOR ENGINE STATIC STRUCTURES

AIRFRAME APPLICATIONS

- DEVELOP TOUGHER MATRIX RESINS

Figure 1

LEWIS PMR POLYMIDE TECHNOLOGY

Studies conducted at the Lewis Research Center led to the development of the concept and class of polyimides known as PMR (for in situ polymerization of monomer reactants) polyimides (refs. 1 and 2). The PMR concept has been adopted by other investigators, and PMR polyimide materials are being offered commercially by the leading suppliers of composite materials. Figure 2 outlines some salient features of the PMR polyimide approach for the fabrication of composites. The reinforcing fibers are hot-melt or solution impregnated with a mixture of monomers dissolved in a low-boiling-point alkyl alcohol. Following fiber impregnation, in situ polymerization of the monomers is caused by heating at temperatures in the range of 250° to 450° F. The final polymerization, an addition reaction, occurs at temperatures in the range of 525° to 660° F without the evolution of undesirable reaction by-products, making it possible to fabricate void-free composite structures. The highly processable polyimides are now making it possible to realize much of the potential of high-temperature polymer matrix composites.

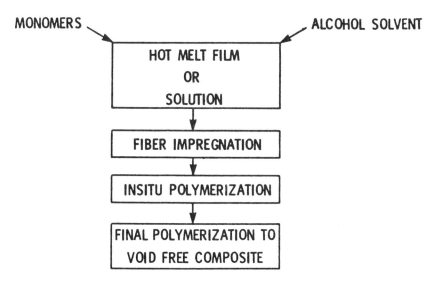

Figure 2

MONOMERS USED FOR PMR-15 POLYIMIDE

The excellent elevated temperature properties and processability of PMR polyimide composites based on the PMR matrix known as PMR-15 have led to their acceptance as viable engineering materials for high-performance structural applications. The structures of the monomers used in PMR-15 are shown in figure 3. The number of moles of each monomer reactant is governed by the following ratio: $2:n:(n + 1)$, where 2, n, and $(n + 1)$ are the number of moles of NE, BTDE, and MDA, respectively. In PMR-15 the value of n is 2.087, corresponding to a formulated molecular weight of 1500. This PMR composition was found to provide the best overall balance of processing characteristics and thermo-oxidative stability at 600° F (ref. 3). Solutions having solids contents in the range of 50 to 85 percent are prepared by simply dissolving the monomer reactants in an alcohol such as methanol. Higher solids content solutions are used for psuedo-hot-melt fiber impregnation techniques.

STRUCTURE	NAME	ABBREVIATION
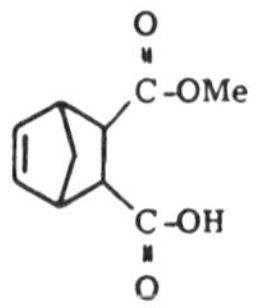 (norbornene monomethyl ester structure)	MONOMETHYL ESTER OF 5-NORBORNENE-2,3-DICARBOXYLIC ACID	NE
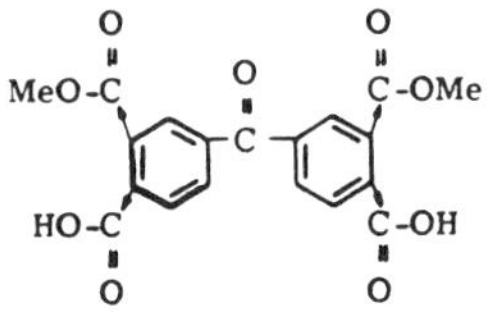 (benzophenone tetracarboxylic dimethyl ester structure)	DIMETHYL ESTER OF 3,3',4,4'-BENZOPHENONETETRACARBOXYLIC ACID	BTDE
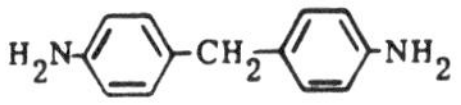 H_2N–⌬–CH_2–⌬–NH_2	4,4'-METHYLENEDIANILINE	MDA

Figure 3

VERSATILITY OF PMR APPROACH

The early studies (ref. 1 and 3) conducted at Lewis clearly demonstrated the versatility of the PMR approach. By varying either the chemical composition of the monomers or the monomer stoichiometry, or both, PMR matrices having a broad range of processing characteristics and properties can be readily synthesized. A PMR composition, designated as PMR-II, has been identified which exhibited improved thermo-oxidative stability compared to PMR-15 (ref. 4). PMR-II has not been accepted as a matrix material because of the lack of a commercial source for one of the monomer reactants used in formulating the resin. A modified PMR-15, called LARC-160, has been developed by substituting an aromatic polyamine for MDA (ref. 5). Other studies (ref. 6) demonstrated the feasibility of using the PMR approach "to tailor make" matrix resins with specific properties. For example, as shown in figure 4, the resin flow characteristics (based on weight of resin flash formed during molding) of PMR polyimides can be varied, or "tailored", over a broad range by simply varying the formulated molecular weight. The higher flow formulations did exhibit decreased thermo-oxidative stability at 550° F.

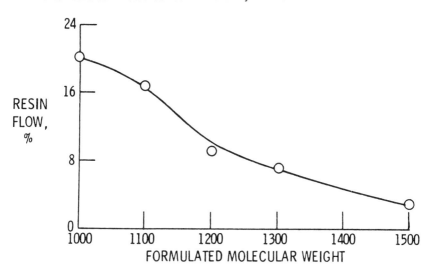

Figure 4

PMR-15 POLYIMIDE MODIFICATIONS FOR IMPROVED PREPREG TACK

Current-technology PMR-15 polyimide prepreg solutions are generally prepared by dissolving the monomer mixture in methanol. Although the volatility of methanol is highly desirable for obtaining void-free composites, it does limit the tack and drape retention characteristics of unprotected prepreg exposed to the ambient. PMR-15 monomer reactants and a mixed solvent have been identified which provide prepreg materials with improved tack and drape retention characteristics without changing the basic cure chemistry or processability (ref. 7). The modifications consist of substituting higher alkyl esters for the methyl esters in NE and BTDE and using a solvent mixture (3:1 methanol/1-propanol) in lieu of pure methanol. As can be seen in figure 5 (left), the ester and solvent modifications extend the tack retention of PMR-15 prepreg to beyond 12 days under ambient conditions. Figure 5 (right) shown that the 600° F interlaminar shear strength (ILSS) properties of Celion 6000 graphite fiber composites made with the modified PMR-15 system are identical to the 600° F (ILSS) properties of composites made with the control, or unmodified, system. The improved tack PMR-15 system should facilitate the fabrication of large complex structures which require long layup times and provide cost savings because of reduced material scrap.

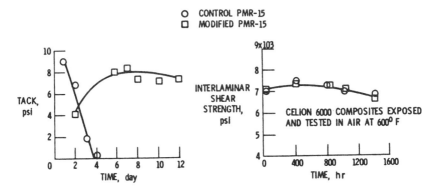

Figure 5

LOWER-CURING-TEMPERATURE PMR POLYIMIDES

The recommended temperature for final cure of PMR-15 is 600° F. This temperature exceeds the temperature capabilities of many industrial autoclave facilities which were originally acquired for curing of epoxy matrix composites. Recent studies (ref. 8) have shown that a significant reduction in cure temperature of PMR-15 can be achieved by replacing 50 mole percent of the NE with p-aminostyrene. Cure studies of the modified system, designated PMR-NV, showed that final cure temperature of PMR-15 could be reduced to 500° F (figure 6, upper left) without sacrificing its 600° F thermo-oxidative stability (figure 6, upper right). As can be seen in figure 6 (bottom), the 600° F interlaminar shear strength (ILSS) properties of Celion 6000 composites made with the PMR-NV matrix are equivalent to the ILSS properties of Celion 6000/PMR-15 composites. The lower cure temperature of the PMR-NV system should be more compatible with existing autoclave facilities and should lead to wider usage of PMR materials.

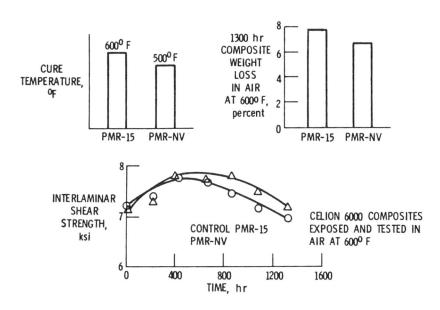

Figure 6

IMPROVED CELION 6000/PMR-15 COMPOSITES

Our continuing research with PMR polyimides has identified a fourth monomer reactant which improved the thermo-oxidative stability and resin flow during composite fabrication (ref. 9). Figure 7 (left) shows the 1500-hour composite weight loss in air at 600° F of Celion 6000 composites made with PMR-15 containing 0 to 20 mole percent N-phenylnadimide (PN). It can be seen that PN levels of 4 and 9 mole percent resulted in improved 600° F thermo-oxidative stability. Figure 7 (right) shows the variation of resin flow during composite processing as a function of PN content. It can be seen that increased resin flow results from the addition of PN. Although the PN-modified PMR-15 composites exhibited lower initial properties at 600° F than unmodified PMR-15 composites, the addition of PN appears to be a promising approach to improve the thermo-oxidative stability of PMR-15.

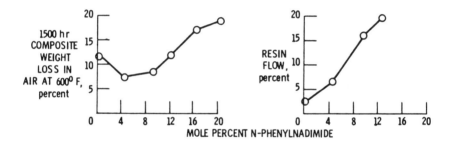

Figure 7

IMPROVED SHEAR STRAIN OF IMIDE-MODIFIED EPOXY

The approach of introducing imide groups into the molecular structure of epoxy resins by reacting epoxy oligomers with novel bis(imide-amine) curing agents was developed by Lewis investigators as a means of improving the thermal characteristics of epoxies (ref. 10). Studies are presently under way at United Technologies Research Center, under contract to NASA, to establish the potential of imide-modified epoxies for improving composite toughness. Figure 8 (left) compares the 10-degree off-axis tensile strengths (ref. 11) of Celion 6000 composite made with a conventional epoxy (A) and an imide-modified epoxy (B). Figure 8 (right) compares calculated resin shear strain for the two composite systems. The resin shear strains were calculated from data generated using 10-degree off-axis tensile tests. It can be seen that the imide-modified resin exhibited more than a two-fold increase in calculated shear strain compared to the conventional epoxy. Studies are currently in progress to establish the correlation between composite toughness and resin shear strain calculated from 10-degree off-axis tensile tests.

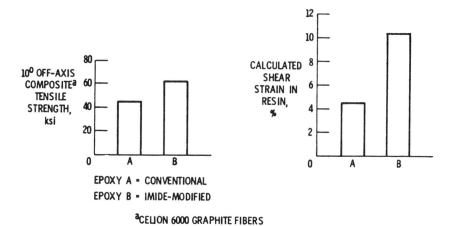

Figure 8

COMPOSITE DURABILITY

Research is being conducted at the Lewis Research Center to develop methodology to predict the life and/or durability of composite structural components in engine service environments. Service environments of major concern are various combinations of temperature, moisture, and mechanical loads. A "generalized" predictive model for predicting the life/durability of graphite fiber/resin matrix composites has been developed (ref. 12). Figure 9 compares experimental data (individual data points) and data predicted by the "generalized" model (parallel lines) for AS graphite/epoxy composites subjected to compression-compression fatigue at room temperature under dry or wet conditions. The important point to note is that the measured data at fracture is above the predicted lines. This indicates the conservative nature of the "generalized" predictive model and should provide credence for its use in preliminary designs.

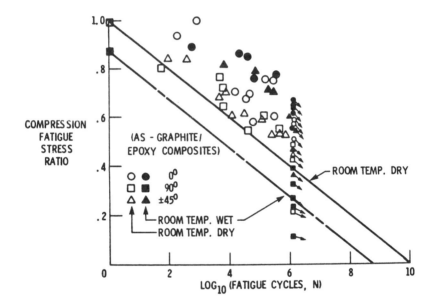

Figure 9

APPLICATIONS OF PMR-15 POLYIMIDE COMPOSITES

One of the most rewarding aspects of the PMR polyimide development has been the successful demonstration of PMR-15 polyimide composite materials as viable engineering materials. Prepregs, molding compounds and even adhesives based on PMR-15 have been commercially available from the major suppliers of composite materials since the mid-seventies. Because of their commerical availability, processability, and excellent retention of properties at elevated temperatures, PMR-15 composites have been used to fabricate a variety of structural components. These components range from small compression-molded bearings to large autoclave-molded aircraft engine cowls and ducts. Processing technology and baseline materials data are being developed for the application of PMR-15 composites in aircraft engines, space structures, and weapon systems. Some representative applications of PMR-15 composites are listed in figure 10. None of the components listed in the figure, with the exception of the ion engine beam shield, are applications in the sense that the components are currently being produced. However, several of the components listed in figure 10 are scheduled for production introduction in the near future. A brief discussion of each of the components listed in figure 10 follows.

COMPONENT	AGENCY	COMPANY
ULTRA-HIGH TIP SPEED FAN FLADES	NASA-LeRC	PWA/TRW
QCSEE INNER COWL	NASA-LeRC	GE
F404 OUTER DUCT	NAVY/NASA-LeRC	GE
F101 DFE INNER DUCT	AIR FORCE	GE
T700 SWIRL FRAME	ARMY	GE
JT8D REVERSER STANG FAIRING	NASA-LeRC	McDONNELL-DOUGLAS
EXTERNAL NOZZLE FLAPS		
PW1120	----	PWA[a]
PW1130	AIR FORCE	PWA
SHUTTLE ORBITER AFT BODY FLAP	NASA-LaRC	BOEING
ION THRUSTER BEAM SHIELD	NASA-LeRC	HUGHES

[a] COMPANY FUNDED

Figure 10

ULTRA-HIGH-TIP-SPEED FAN BLADES

The blade illustrated in figure 11 was the first structural component fabricated with a PMR-15 composite material. The reinforcement is HTS graphite fiber. The blade design was conceived by Pratt & Whitney Aircraft (PWA) for an ultra-high-speed fan stage (ref. 13). Blade tooling and fabrication were performed by TRW Equipment (ref. 14). The blade span is 11 in. and the chord is 8 in. The blade thickness ranges from about 0.5 in. just above the midpoint of the wedge-shaped root to 0.022 in. at the leading edge. At its thickest section the composite structure consists of 77 plies of material arranged in varying fiber orientation. The "line of demarkation" visible at approximately one-third the blade span from the blade tip resulted from a required change in fiber orientation from 40 degrees in the lower region to 75 degrees in the upper region to meet the torsional stiffness requirements. Ultrasonic and radiographic examination of the compression-molded blades indicated that they were defect free. Although some minor internal defects were induced in the blades during low-cycle and high-cycle fatigue testing, the successful fabrication of these highly complex blades established the credibility of PMR-15 as a processable matrix resin.

Figure 11

APPLICATION OF COMPOSITES ON QUIET CLEAN SHORT-HAUL EXPERIMENTAL ENGINE (QCSEE)

The Quiet, Clean, Short-Haul Experimental Engine (QCSEE) program was initiated to develop a propulsion technology base for future powered-lift short-haul aircraft. One of the major areas of new technology investigated under the QCSEE program was the application of advanced composite materials to major engine hardware. Figure 12 shows a cutaway drawing of the under-the-wing (UTW) QCSEE engine. Composite materials were used for fan blades, the fan frame, and nacelle components. The blades, frame and all nacelle components with the exception of the inner cowl were fabricated from Kevlar or graphite fibers in an epoxy matrix resin. The inner cowl was made of graphite fibers in PMR-15 polyimide.

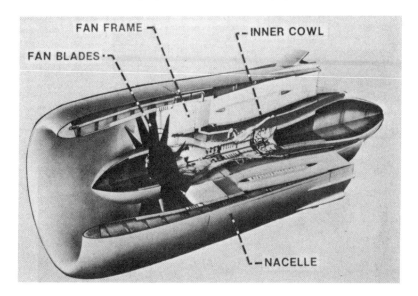

Figure 12

GRAPHITE FIBER/PMR-15 QCSEE INNER COWL

Figure 13 shows the composite inner cowl installed on the UTW QCSEE engine developed by General Electric (GE) under contract with the Lewis Research Center (ref. 15). The cowl defines the inner boundary of the fan air flowpath from the fan frame to the engine core nozzle. The cowl was autoclave-fabricated by GE from PMR-15 and T300 graphite fabric. The cowl has a maximum diameter of about 36 in. and is primarily of honeycomb sandwich construction. HRH327 fiberglass polyimide honeycomb was used as the core material. Complete details about the cowl fabrication process are given in reference 16. The cowl was installed on the QCSEE engine and did not exhibit any degradation after more than 300 hours of ground engine testing. The maximum temperature experienced by the cowl during testing was 500° F (ref. 17). The successful autoclave fabrication and ground engine test results of the QCSEE inner cowl established the feasibility of using PMR-15 composite materials for large engine static structures.

Figure 13

Selected Additional Subjects 363

GENERAL ELECTRIC F404-GE-400 TURBOFAN

Under a jointly sponsored U. S. Navy/NASA Lewis program (NAS3-21854), GE is developing a T300 graphite fabric/PMR-15 composite outer duct to replace the titanium duct presently used on the F404 engine, shown in figure 14, for the Navy's F18 strike fighter. The titanium duct (the waffle-like structure) is a sophisticated part made by forming and machining titanium plates followed by chem-milling to reduce weight. A preliminary cost-benefit study indicated that significant cost and weight savings (ref. 18) could be achieved by replacing the titanium duct with a composite duct.

Figure 14

364 Tough Composite Materials

GRAPHITE FIBER/PMR-15 POLYIMIDE OUTER DUCT FOR GE F404 ENGINE

Figure 15 is a photograph of the full-scale composite duct (38 in. diameter x 65 in. length x 0.080 in. wall thickness) that was autoclave-fabricated from T300 graphite fabric and PMR-15 polyimide. The duct was proof pressure checked successfully, prior to ground engine testing, to 108 psi (150 percent of operating pressure). The F404 composite outer duct differs from the QCSEE inner cowl in several important respects. The F404 duct is a monolithic composite structure and needs to withstand fairly high loads, and perhaps most importantly, the F404 duct is to be a production component and not a "one-of-a-kind" demonstration component.

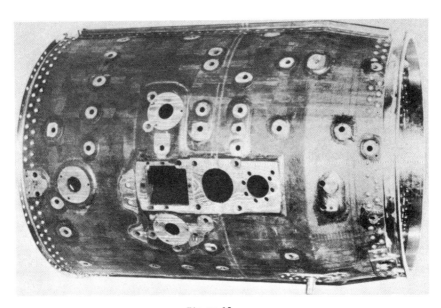

Figure 15

GRAPHITE FIBER/PMR-15 POLYIMIDE OUTER DUCT ON GE F404 ENGINE

Figure 16 shows the composite duct installed on an F404 test engine. The composite duct has successfully withstood over 1000 accelerated mission test cycles during a total engine exposure time of 700 hours. The graphite/PMR-15 polyimide composite duct is scheduled for production introduction in 1985.

Figure 16

SCHEMATIC OF GE T700 PMR-15 COMPOSITE SWIRL FRAME

The current bill-of-materials inlet particle separator swirl frame on GE's T700 engine is an all-metal part that involves machining, shape-forming, welding, and brazing operations. Design studies conducted under U. S. Army contract number DDAK51-79-C-0018 indicated that the fabrication of a metal/composite swirl frame could result in cost and weight savings of about 30 percent. Figure 17 shows a schematic diagram of a section of the metal/composite swirl frame that was fabricated from 410 stainless steel and various kinds of PMR-15 composite materials. The outer casing uses stainless steel in the flow path area to meet anti-icing temperature requirements and T300 and glass fabric/PMR-15 hybrid composite to meet structural requirements. The T300/glass hybrid composite was selected on the basis of both cost and structural considerations. An aluminum-coated glass fabric PMR-15 composite material is utilized in the inner-hub flowpath to meet heat transfer requirements for anti-icing. The glass fabric/PMR-15 composite utilized for the front edge and front inner surface was selected because of cost as well as temperature considerations. A full-scale (O.D. ~ 20 in.) metal/composite swirl frame has been subjected to sand erosion and ice ball impact tests. The metal/composite swirl frame provided improved particle separation and successfully met the impact test requirements. Fabrication feasibility has been demonstrated, and if the metal/composite swirl frame successfully meets all of the performance requirements, the metal/composite T700 swirl frame will be introduced into production in 1985.

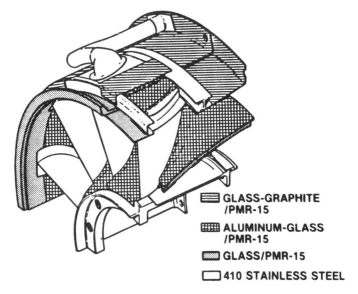

Figure 17

APPLICATIONS OF GRAPHITE FIBER/PMR-15 POLYIMIDE COMPOSITES ON PW1120 ENGINE

Figure 18 is a schematic showing "committed" and "possible" applications of graphite/PMR-15 composite materials on the PW1120 turbojet currently being developed by Pratt & Whitney Aircraft/Government Products Division (PWA/GPD). A committed application is an application for which a metal back-up component is not being developed. The committed applications for graphite/PMR-15 composites on the PW1120 at this time are the external nozzle flaps and the airframe interface ring. PWA/GPD is in the process of completing its assessment of the various "possible" applications and anticipates that many of these will also become "committed", if engine test schedules can be met. The PW1120 engine is currently scheduled for production deliveries in 1986. Graphite/PMR-15 external nozzle flaps have been committed for production by PWA/GPD for its PW1130 turbofan engine. Production deliveries of the PW1130 are scheduled for 1984. Prepregs made from T300 or Celion 3000 uniweave fabrics and PMR-15 are being evaluated for fabrication of the nozzle flaps used on both the PW1120 and PW1130 engines.

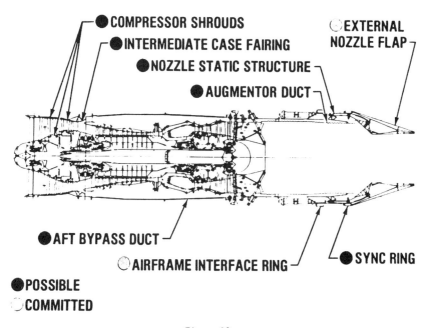

Figure 18

DC-9 DRAG REDUCTION

Figure 19 shows a photograph of a DC-9. The inserts schematically depict the design of the presently used metal reverser stang fairing and a composite redesigned fairing developed by Douglas Aircraft Company under the NASA Lewis Engine Component Improvement Program (ref. 19). Studies had shown that a redesigned fairing provided an opportunity to reduce baseline drag and would result in reduced fuel consumption. The fairing serves as the aft enclosure for the thrust reverser actuator system on the nacelle of the JT8D and is subjected to an exhaust temperature of 500° F during thrust reversal. A Kevlar fabric/PMR-15 composites fairing has been autoclave fabricated and flight tested. Compared to the metal component, the composite fairing resulted in a one percent airplane drag reduction (1/2 percent had been anticipated) and a 40-percent reduction in component weight.

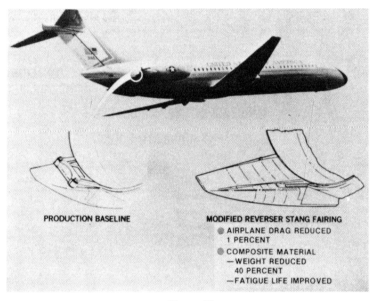

Figure 19

KEVLAR FABRIC/PMR-15 REVERSER STANG FAIRING

Figure 20 shows a photograph of the Kevlar fabric/PMR-15 reverser stang fairing. The weight of the composite fairing was found to be 40 percent less than the calculated weight of a fairing of the same shape made from aluminum.

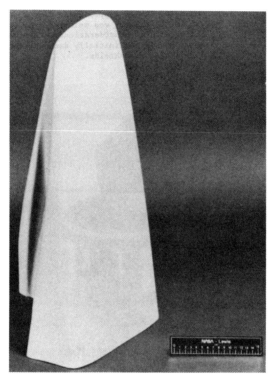

Figure 20

GLASS FABRIC/PMR-15 BEAM SHIELD INSTALLED ON MERCURY ION THRUSTER

Figure 21 shows a mercury ion thruster for an auxiliary propusion system being built by Hughes Space and Communications Group under contract to NASA Lewis. The ion propulsion system is scheduled for launch and testing on a future Shuttle flight. The thruster is equipped with a glass fabric/PMR-15 composite beam shield to protect the solar cell arrays and sensitive instrumentaion on the spacecraft from ion-beam damage. The composite shield (approximate dimensions: 10 in. diameter by 8 in. length by 0.040 in. thickness) was selected over tantalum and titanium because of weight and structural considerations. The feasibility of using a glass fabric/PMR-15 composites shield was initially demonstrated by in-house fabrication and testing of full-scale beam shields.

Figure 21

CONCLUDING REMARKS

The in situ polymerization of monomer reactants (PMR) approach has been demonstrated to be a powerful approach for solving many of the processing difficulties associated with the use of high-temperature resistant polymers as matrix resins in high-performance composites. PMR-15, the PMR polyimide discovered in the early seventies, provides the best overall balance of processing characteristics and elevated temperature properties. The excellent properties and commercial availability of composite materials based on PMR-15 have led to their acceptance as viable engineering materials. PMR-15 composites are currently being used to produce a variety of high-quality structural components. Increased use of these materials is anticipated in the future.

PMR-15 POLYIMIDE COMPOSITES:

- **PROVIDE EXCELLENT PROCESSABILITY**
- **PROVIDE EXCELLENT HIGH TEMPERATURE PROPERTIES**
- **BEING ACCEPTED AS VIABLE ENGINEERING MATERIALS**
- **BEING USED TO FABRICATE HIGH QUALITY STRUCTURAL COMPONENTS**

Figure 22

REFERENCES

1. Serafini, T. T.; Delvigs, P.; and Lightsey, G. R.: Thermally Stable Polyimides from Solutions of Monomeric Reactants, J. Appl. Polym. Sci., Vol. 16, 1972. pp. 905-915.

2. Serafini, T. T.; Delvigs, P.; and Lightsey, G. R.: Preparation of Polyimides from Mixtures of Monomeric Diamines and Esters of Polycarboxylic Acids, U.S. Patent 3,745,149, July 10, 1973.

3. Delvigs, P.; Serafini, T. T.; and Lightsey, G. R.: Addition-Type Polyimides from Solutions of Monomeric Reactants, NASA TN D-6877, 1972. (Also Proc. of 17th SAMPE National Symposium and Exhibition, April 1972.)

4. Serafini, T. T.; Vannucci, R. D.; and Alston, W. B.: Second Generation PMR Polyimides, NASA TM X-71894, 1976.

5. St. Clair, T. L. and Jewell, R. A.: Solventless LARC-160 Polyimide Matrix Resin. Proc. of 23rd SAMPE National Symposium and Exhibition, May 1978.

6. Serafini, T. T. and Vannucci, R. D.: Tailor Making High Performance Graphite Fiber Reinforced PMR Polyimides, NASA TM X-71616, 1975. (Also Proc. of 30th SPI RP/Composites Institute Conference, February 1975.)

7. Vannucci, R. D.: PMR Polyimides Modifications for Improved Prepreg Tack, NASA TM-82951, 1982.

8. Delvigs, P.: Lower-Curing-Temperature PMR Polyimides, NASA TM-82958, 1982.

9. Pater, R. H.: Novel Improved PMR Polyimides, NASA TM-82733, 1981. (Also Proc. of 13th SAMPE National Technical Conference, October 1981.)

10. Serafini, T. T.; Delvigs, P.; and Vannucci, R. D.: Curing Agent for Polyepoxides and Epoxy Resins and Composites Cured Therewith, U.S. Patent 4,244,857, January 13, 1981.

11. Chamis, C. C. and Sinclair, J. H.: Ten-Deg Off-Axis Test for Shear Properties in Fiber Composites, Experimental Mechanics, Vol. 17, No. 9, 1977, pp. 339-346.

12. Chamis, C. C. and Sinclair, J. H.: Durability/Life of Fiber Composites in Hygrothermomechanical Environments, Composite Materials: Testing and Design, Sixth Conference, ASTM STP 787, I. M. Daniel, Ed., ASTM 1982, pp. 498-512.

13. Halle, J. E.; Burger, E. D.; and Dundas, R. E.: Ultra-High Speed Fan Stage with Composite Rotor, NASA CR-135122, 1977.

14. Cavano, P. J.: Resin/Graphite Fiber Composites, NASA CR-134727, 1974.

15. Adamson, A. P.: Proc. of Conference on Quiet Powered-Lift Propulsion, NASA CP-2077, 1978, pp. 17-29.

16. Ruggles, C. L.: QCSEE Under-the-Wing Graphite/PMR Cowl Development, NASA CR-135279, 1978.

17. Stotler, C. L.: Proc. of Conference on Quiet Powered-Lift Propulsion, NASA CP-2077, pp. 83-109.

18. Stotler, C. L.: Development Program for a Graphite/PMR-15 Polyimide Duct for the F404 Engine, Proc. of 25th SAMPE National Symposium and Exhibition, May 1980, pp. 176-187.

19. Kawai, R. T. and Hrach, F. J.: Development of a Kevlar/PMR-15 Reduced Drag DC-9 Nacelle Fairing, AIAA-80-1194, AIAA/SAE/ASME 16th Joint Propulsion Conference, June 1980.

FATIGUE AND FRACTURE RESEARCH IN COMPOSITE MATERIALS

T. Kevin O'Brien

NASA Langley Research Center
Hampton, VA

INTRODUCTION

NASA Langley has a major research program to understand and characterize the fatigue, fracture, and impact behavior of composite materials (fig. 1). Bolted and bonded joints are included. The scope of this work is generic so that the solutions developed will be useful for a wide variety of structural applications. The analytical tools developed are used to demonstrate the damage tolerance, impact resistance, and useful fatigue life of structural composite components. Furthermore, much recent emphasis is on developing and analyzing standard tests for screening improvements in materials and constituents.

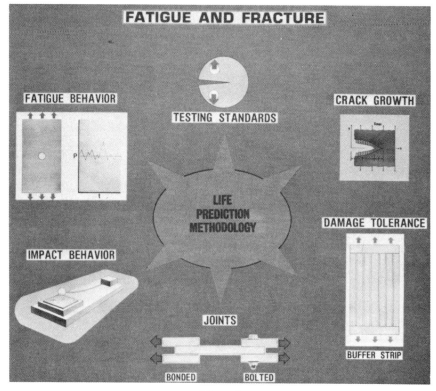

Figure 1

A UNIFYING STRAIN CRITERION TO PREDICT FRACTURE TOUGHNESS OF COMPOSITE LAMINATES

In references 1 and 2 a method was developed to predict composite fracture toughness from fiber and matrix properties. A laminate was assumed to fail when the strains at the crack tips in the principal load-carrying plies reach a critical level, regardless of layup. (See fig. 2.) The singular term in a series representation of the orthotropic strain field was chosen to represent the fiber strains. The fibers in the principal load-carrying plies were oriented at an angle θ to the applied load. The singular coefficient Q_c is given by $Q_c = K_Q \xi / E_y$, where K_Q is the usual stress intensity factor, E_y is the laminate modulus in the loading direction, and ξ is a function that depends only on the laminate elastic constants and the angle of the principal load-carrying plies. A critical level of strains at failure implies a constant value of Q_c, independent of ply orientations. Furthermore, Q_c is proportional to the ultimate tensile failing strain of the fibers, independent of the matrix material. (See fig. 2.) Therefore, the fracture toughness of all fibrous composite laminates can be computed from the ultimate tensile failing strain of the fibers and the elastic constants.

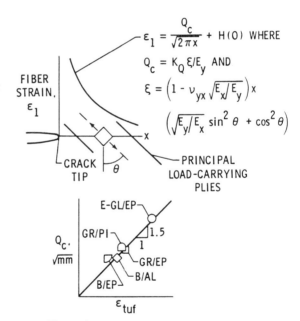

Figure 2

MEASURED AND PREDICTED VALUES OF FRACTURE TOUGHNESS

Figure 3 shows predicted and measured values of composite fracture toughness for a number of different layups and materials. The value of K_Q increases in proportion to fiber strength. Hence, the laminates with boron fibers have the highest K_Q and those with E-glass the lowest. The predictions, which are usually within 10 percent of the measurements, are quite good. The discrepancies between predicted and measured values of K_Q may be due largely to matrix and fiber failures at the crack tips that occur before overall failure. These crack tip failures, which alter the local fiber stresses, are not presently accounted for in the analysis. Micromechanical analyses are being developed that do account for these failures. In addition, experiments are being conducted to measure the type and extent of crack tip failures for various layups.

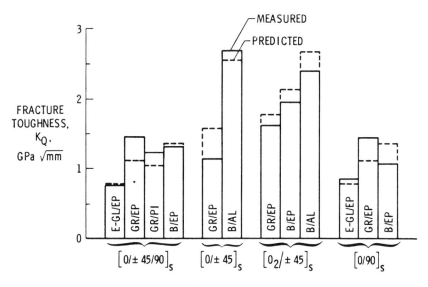

Figure 3

IMPROVING DAMAGE TOLERANCE OF COMPOSITE LAMINATES
WITH BUFFER STRIPS - TEST RESULTS

Buffer strips can greatly increase the tensile strength of damaged graphite/epoxy laminates while increasing weight only a little. In addition, recent work has shown that S-glass buffer strips can be woven into the 0° graphite plies with little additional expense (ref. 3). Figure 4 shows the remote strain for three quasi-isotropic panels containing center slits and S-glass buffer strips. Fracture initiates in panels with (filled symbols) and without (dashed curve) buffer strips at the same remote strain. However, the buffer strips arrest the fractures because they have a higher modulus of resilience than the graphite. The eventual failing strains of the panels are more than two times the strain at which the fracture initiated from the longest slit. Thus, in a damage tolerance situation, the strengths of these panels can be twice that of plain laminates.

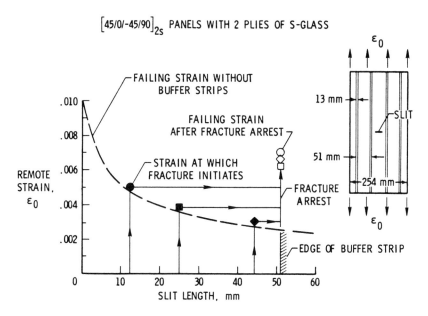

Figure 4

IMPROVING DAMAGE TOLERANCE OF COMPOSITE LAMINATES

WITH BUFFER STRIPS - ANALYSIS

In addition to the buffer strip panels in figure 4, panels were also made with other buffer materials, various numbers of buffer plies and spacings, and various layups (ref. 3). Figure 5 shows the ratios of cracked to uncracked strength for these tests plotted against the buffer strip spacing (which is the arrested crack length) multiplied by the ratios of number of 0° plies and modulus of resilience for graphite to buffer material. Each symbol represents an average of three tests. Round symbols represent S-glass buffer strips; square symbols represent Kevlar buffer strips. Filled and open symbols differentiate between the two layups (shown in the key) used for the main panel material. A shear-lag analysis indicates that for large values of the abscissa, the strength ratio should vary inversely with the square root of the abscissa. S-glass has the highest modulus of resilience, and hence gives the highest strengths. For small values of the abscissa, the strengths are limited to the ultimate net section strength, which is 75 percent of the uncracked strength. (The buffer strip spacing, W_a, is 25 percent of the panel width.) The test data agree very well with this relationship. Additional tests are currently being conducted for a wider range of buffer strip parameters and for woven buffer material. Tests are also being conducted to determine the influence of environment and fatigue loading on strengths. Furthermore, analyses are being developed to better predict the strengths in terms of the buffer strip parameters.

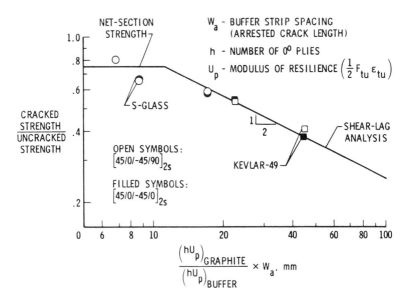

Figure 5

HOLE ELONGATION UNDER CYCLIC LOADING
IN BOLTED COMPOSITE JOINTS

Bolt holes in composite joints can elongate under cyclic loading and thereby reduce joint stiffness. Single-fastener joints were tested in reference 4 using different combinations of tension-tension fatigue loading and bolt clampup torques. Figure 6 shows measured hole elongations from three tests that produced about the same fatigue life. This figure shows the influence of clampup on hole elongation. For the case of pin bearing with no clampup constraint, the figure shows that the bolt hole did not elongate before it failed in fatigue. But with "finger-tight" clampup and a higher cyclic stress, the hole elongated extensively before it failed. Much less elongation is shown for the third case, which has a typical 2.82 N·m clampup for the 6.35-mm-diameter bolt. Other tests were conducted to measure hole elongation and strength under static loading. Future research will focus on analyzing damage mechanisms for bearing-loaded holes.

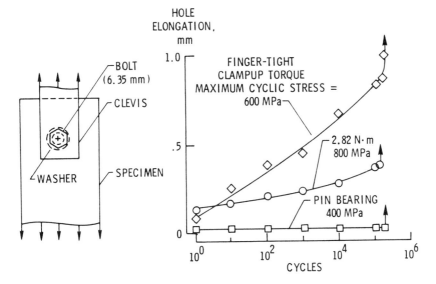

Figure 6

BOLT CLAMPUP RELAXATION

Recent studies have shown that bolt clampup improves the strength of composite joints. This improvement, however, may decrease somewhat if the bolt clampup force relaxes during long-term exposure. A viscoelastic relaxation analysis has been developed (ref. 5) for steady-state exposure conditions. A simple double-lap bolted joint was analyzed using a finite-element procedure. Typical results are shown as symbols in the accompanying figure 7. For the room-temperature-dry (RTD) reference case, relaxations of 8, 13, 20, and 30 percent were calculated for 1 day, 1 week, 1 year, and 20 years, respectively. As expected, moisture increased the clampup relaxation compared to the RTD case. Reference 5 also includes results for elevated-temperature exposures. In addition to the finite-element analyses, a simple empirical equation was developed to calculate clampup relaxation. First, this equation was fitted (dashed curve) to the RTD finite-element results to establish two constants (F_1 and n). Then the equation was generalized to account for moisture by substituting viscoelastic shift factors a_{TH} from the literature. The solid curves show that this equation agrees quite closely with the finite-element results. Recently, this equation was extended to account for transient environments (ref. 6).

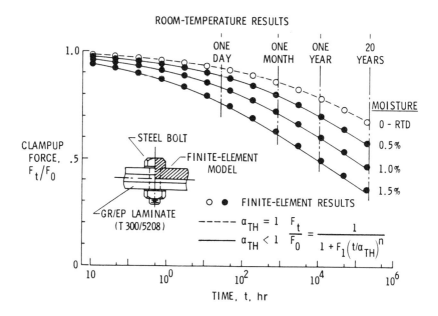

Figure 7

3-D STRESSES AT LAMINATE BOLT HOLES

High interlaminar stresses develop near holes in composite laminates. These interlaminar stresses were analyzed in reference 7 for a [0/90]$_s$ graphite/epoxy laminate with a circular hole and remote uniaxial loading. These stresses were calculated using a 3-D finite-element model with a very high mesh refinement near the ply interface. Typical stress results are presented in figure 8, and for convenience they are normalized by the remote stress S_g. The two curves in this figure represent the interlaminar normal and shear stresses, σ_z and $\sigma_{z\theta}$, respectively, along the hole boundary. These stresses provide a basis for locating delaminations at laminate holes. The 3-D finite-element procedures are currently being extended to calculate interlaminar stresses due to thermal and hygroscopic effects.

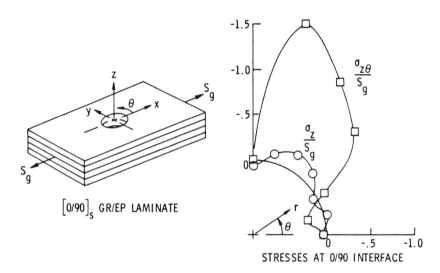

Figure 8

DAMAGE THRESHOLD UNDER LOW-VELOCITY IMPACT OF COMPOSITES

Low-velocity impact can easily damage composite laminates by delaminating bonded layers or breaking fibers. Impact damage was studied to analyze the impact mechanics, determine failure criteria, and develop a simple impact test method. The test device consists of a hardened 1-inch-diameter steel ball at the end of a cantilevered rod (fig. 9). The ball strikes a 4-inch-square composite plate that is bonded to an aluminum support plate. The support plate has a circular aperture ranging in diameter from 1/2 to 3-1/2 inches. The ball and laminate are wired so that while they are in contact a circuit is completed and contact time is recorded on an oscilloscope. The expected impact duration, maximum impact force, and maximum shear and flexural stresses are calculated from an analysis based on the static first-mode deformations. The predicted durations and the test results agree within ±3 percent. The criteria for failure were found to be matrix ultimate shear stress and fiber ultimate stress. The right-hand figure shows that the analysis correctly predicts the damage threshold within about ±10 percent.

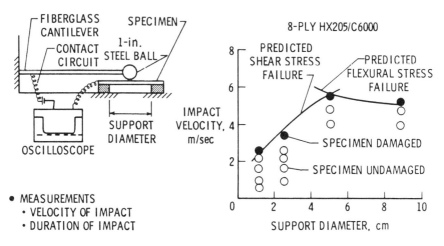

- MEASUREMENTS
  - VELOCITY OF IMPACT
  - DURATION OF IMPACT

- MODEL
  - DYNAMIC ANALYSIS OF DURATION, FORCE, ETC.
  - ONLY FIRST-MODE DEFORMATIONS CONSIDERED

- MODEL FAILURE PREDICTION AGREES WITH EXPERIMENT TO WITHIN ±10%

Figure 9

COMPOSITE IMPACT SCREENING BY STATIC INDENTATION TESTS

Graphite/epoxy laminates can lose up to 50 percent of their original strength from low-velocity impact damage. For laminates up to at least 16 plies, a static indentation test on a specimen bonded between two aperture plates yields failures almost identical to the failures from an impact (ref. 8). A static test, however, allows sufficient time for analysis of the failure sequence (fig. 10). The load deflection curve for an undamaged plate with a small span is shown in the left figure; the flexural stiffness from the shear strength of the matrix transfers most of the load to the supports. Damage progression occurs as matrix shear failure causes delaminations to grow, until only a membrane of delaminated fibers supports the load. Then membrane failure controls penetration load. For this configuration, matrix shear strength dominates the low-level impact damage, and the extent of delamination is limited by the fiber toughness. The figure on the right indicates that for larger aperture plates, failure is dominated by the failure energy of the overall membrane. Matrix failure, in this case, releases little energy; failure here is governed predominantly by fiber strength.

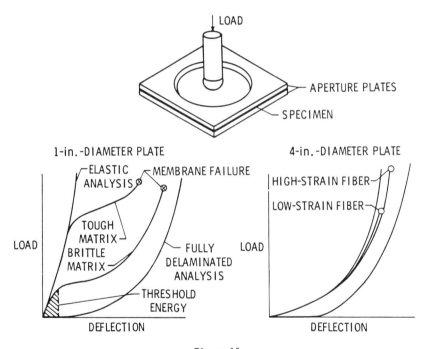

Figure 10

STRAIN ENERGY RELEASE RATE FOR DELAMINATION GROWTH DETERMINED

A simple expression was derived for the strain energy release rate, G, for edge delamination growth in an unnotched laminate (ref. 9). This simple expression shown in figure 11 has several advantages. First, G is independent of delamination size. It depends only on the applied strain, ε, the specimen thickness, t, the stiffness of the undamaged laminate, E_{LAM}, and the stiffness of the laminate when it was completely delaminated, E^*. Second, E_{LAM} and E^* can be calculated from simple laminated-plate theory. Third, because E^* depends on the location of the delaminated interface(s), E^*, and hence G, are sensitive to the location of damage in the laminate thickness. The simple equation was used to develop criteria to predict the onset and growth of delaminations in realistic, unnotched laminates.

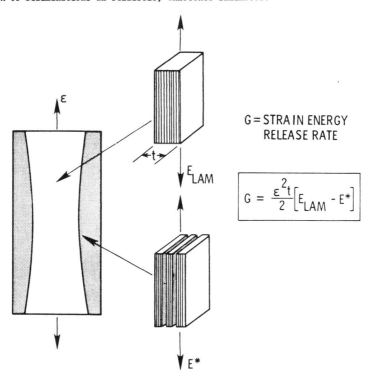

Figure 11

DELAMINATION ONSET PREDICTED

To predict the onset of delaminations in realistic unnotched laminates (ref. 9), quasi-static tension tests were conducted on the $[\pm 30/\pm 30/90/\overline{90}]_s$ laminates. The applied strain recorded at the onset of delamination, ε_c, was used to calculate a critical strain energy release rate, G_c. Then, G_c was used to predict the onset of delamination in more complex laminates. Figure 12 shows data and predictions for three $[+45_n/-45_n/0_n/90_n]_s$ laminates all having the same layup but with different ply thicknesses. For example, n = 1 is an 8-ply laminate, n = 2 is a 16-ply laminate, and n = 3 is a 24-ply laminate. The predictions agreed well with experimental data, indicating that G_c was a material property. Furthermore, the trend of lower ε_c for thicker laminates was correctly predicted. This trend could not be predicted by a critical interlaminar stress criterion.

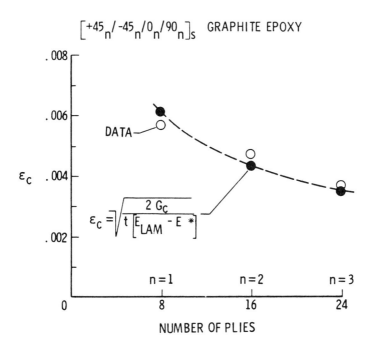

Figure 12

EDGE DELAMINATION TENSION TEST MEASURES
INTERLAMINAR FRACTURE TOUGHNESS

A simple test has been developed for measuring the interlaminar fracture toughness of composites made with toughened matrix resins (ref. 10). The test involves measuring the stiffness, E_{LAM}, and nominal strain at onset of delamination, ε_c, during a tension test of an 11-ply $[\pm30/\pm30/90/\overline{90}]_s$ laminate (fig. 13). These quantities, along with the measured thickness t, are substituted into a closed-form equation for the strain energy release rate, G, for edge delamination growth in an unnotched laminate (ref. 9). The E* term in the equation is the stiffness of the $[\pm30/\pm30/90/\overline{90}]_s$ laminate if the 30/90 interfaces were completely delaminated. It can be calculated from the simple rule of mixtures equation shown in figure 13 by substituting the laminate stiffness measured during tension tests of $[\pm30]_s$ and $[90]_n$ laminates. The critical value of G_c at delamination onset is a measure of the interlaminar fracture toughness of the composite. This edge delamination test is being used by Boeing, Douglas, and Lockheed under the NASA ACEE (Aircraft Energy Efficiency) Key Technologies contracts to screen toughened resin composites for improved delamination resistance (ref. 11).

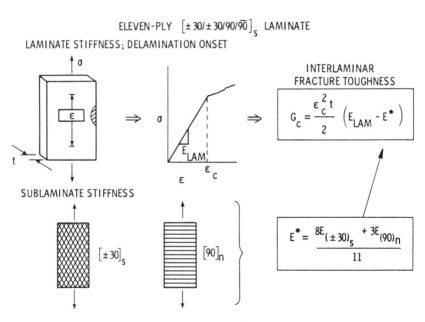

Figure 13

MIXED-MODE STRAIN ENERGY RELEASE RATES DETERMINED

A quasi-three-dimensional finite-element analysis (ref. 12) was performed to determine the relative crack opening (mode I) and shear (mode II) contributions of the $[\pm 30/\pm 30/90/\overline{90}]_s$ edge delamination specimen (ref. 10). Delaminations were modeled in the 30/90 interfaces where they were observed to occur in experiments. Figure 14 indicates that the total G represented by G_I plus G_{II} reaches a value prescribed by the closed-form equation derived from laminated-plate theory and the rule of mixtures. Furthermore, like the total G, the G_I and G_{II} components are also independent of delamination size. In addition, the percentage, G_I/G_{II}, is fixed for a particular layup and does not change significantly for different resin composites as long as the laminates have the same kinds of fibers.

FINITE ELEMENT ANALYSIS OF MIXED MODE PERCENTAGES
$[\pm 30/\pm 30/90/\overline{90}]_s$ LAMINATE

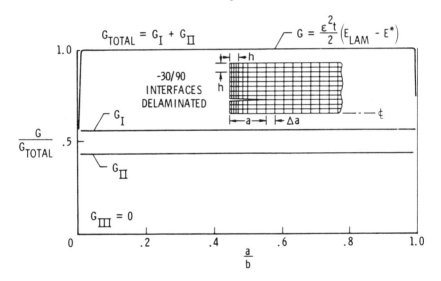

Figure 14

INTERLAMINAR FRACTURE TOUGHNESS OF GRAPHITE COMPOSITES MEASURED

Two test methods are being developed to measure the interlaminar fracture toughness of graphite-reinforced composites (ref. 10). The first is a pure crack opening (mode I) double-cantilever-beam test. The second is the NASA-developed crack opening and shear, mixed-mode (modes I and II), edge delamination tension test (refs. 10 and 11). Figure 15 shows results of these measurements for a relatively brittle 350°F-cure epoxy (5208), a tougher 250°F-cure epoxy (H-205), and a still tougher rubber-toughened 250°F-cure epoxy (F-185). Results indicate that for the brittle epoxy, even in the mixed-mode test, only the crack opening fracture mode contributes to delamination. However, for the tougher 250°F-cure epoxy and its rubber-toughened version, both the crack opening and shear fracture modes contribute to delamination. Hence, although both tests indicate relative improvements among materials, one test alone is not sufficient to quantify interlaminar fracture toughness.

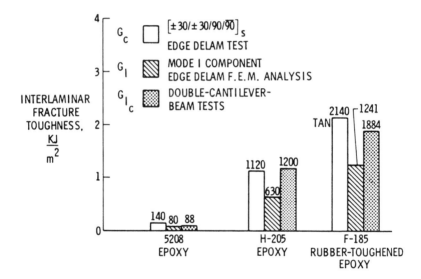

Figure 15

INVESTIGATION OF INSTABILITY-RELATED DELAMINATION GROWTH

Under compression load fatigue, delaminations in composites sometimes induce localized buckling, causing high interlaminar stresses at the ends of the delamination. Rapid delamination growth and loss of structural stability often ensue. Because delamination growth can lead to structural instability, the growth process must be understood. To improve our understanding, through-width delaminations (fig. 16) were studied experimentally and analytically (ref. 13). The figure shows a comparison of measured growth rates and calculated strain energy release rates, which are a measure of the intensity of stresses at the crack tip. In the figure, G_I and G_{II} are the energy release rates related to peel and shear stresses, respectively. Note that both G_I and the growth rate first increase then decrease rapidly with crack extension; G_{II} increases monotonically and does not reflect the change in growth rate. Apparently, the rate of delamination growth is governed by the intensity of the peel stress field. Hence, prediction of crack growth depends on an accurate assessment of the peel stress field around the crack tip. Further work will concentrate on quantifying the relationship between calculated peel stress or related parameters (e.g., G_I) and delamination growth rates.

COMPARISON OF STRAIN-ENERGY-RELEASE RATES AND DELAMINATION GROWTH RATES

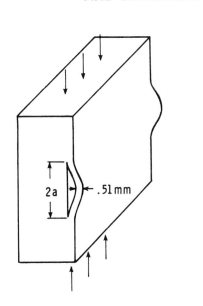
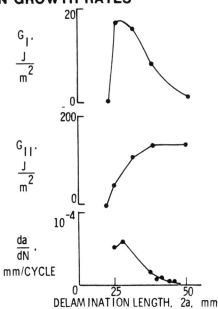

Figure 16

LOCAL DELAMINATION CAUSES TENSILE FATIGUE FAILURES

A study of damage development during tension-tension fatigue loading of unnotched $[\pm 45/0/90]_s$ laminates has demonstrated that local delamination is responsible for fatigue failures at cyclic load levels below the static tensile strength (ref. 14). The circular symbols in figure 17 show the maximum cyclic load, P_{max}, plotted as a function of the number of load-controlled fatigue cycles, N, needed (1) to create delaminations along the edge in 0/90 interfaces (open symbols) and (2) to cause fatigue failures (solid symbols). The arrows extending to the right of data points at or near 10^6 cycles indicate runouts, i.e., no fatigue failures. The square symbols in figure 17 indicate the mean value of (1) load at onset of 0/90 interface delamination along the edge (open symbol) and (2) load at failure (closed symbol) in quasi-static tension tests. During these quasi-static tests, edge delaminations grew almost entirely through the specimen width before failure. Hence, the initial static tensile strength reflects the presence of large 0/90 interface edge delaminations. Yet the endurance limit for fatigue failure was 70 percent of this static tensile strength. Fatigue tests run at cyclic load levels below this limit, but above 40 percent of the tensile strength, contained extensive edge delaminations just like those observed in quasi-static tests. However, specimens in these tests did not fail in fatigue. But tests run at or above the 70-percent endurance limit also developed local delaminations in +45/-45 interfaces. These local delaminations, which originated from matrix cracks in the surface +45° plies, reduced the local cross section and changed the local stiffness. These changes in local stiffness and cross section increased the local strain in the 0° plies, resulting in fiber fracture and laminate failure. Future work will concentrate on predicting fatigue endurance limits using fracture mechanics models of local delamination.

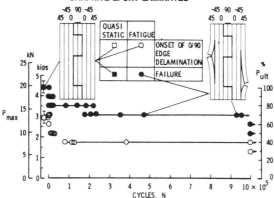

Figure 17

TEST METHODS FOR COMPRESSION FATIGUE OF COMPOSITES

Test methods developed for the fatigue evaluation of airframe metals, primarily aluminum alloys, may not be appropriate for composites. The results of two investigations (ref. 15) aimed at development of test methods for composites loaded primarily in cyclic compression are shown in figure 18. The left side of the figure shows the results of simple cyclic load tests that were conducted to select a method for preventing column buckling in fatigue tests of thin coupons containing an open hole. The test results show that the best configuration is one that does not limit the localized buckling that develops near the hole due to fatigue-induced delaminations. Limiting the localized buckling causes the test to yield an unrealistically long life estimate. The right side of the figure shows the results of simulated flight loading tests that were conducted to determine what parts of the flight load spectrum could be deleted from the simulation to shorten test time without affecting the test result. The results show that deletion of 90 percent of the low loads did not change the test life very much, but that deletion of just a few of the rare, high loads led to very long test lives. Therefore, to be on the safe side in tests on composite structure, the high loads must not be deleted from the test spectrum as would normally be done in tests on aluminum structure.

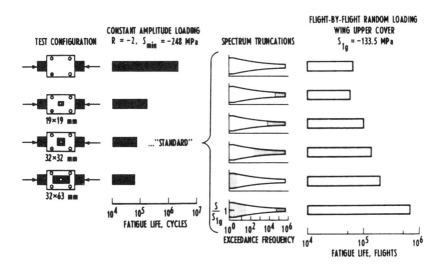

Figure 18

FATIGUE DAMAGE IN BORON/ALUMINUM LAMINATES

Because an aluminum matrix has a modulus 20 times higher than a polymer matrix, matrix cracking in aluminum matrix composites strongly influences laminate stiffness. After long periods of fatigue loading, matrix cracks form in boron/aluminum composites and stiffness may drop significantly (ref. 16). A simple analysis has been developed to predict these laminate stiffness reductions due to fatigue (ref. 17). The analysis is based upon the elastic modulus of the fiber and matrix, fiber volume fraction, fiber orientation, and cyclic-hardened yield stress of the matrix material. It readily predicts the laminate secant modulus of the composite at the stabilized damage state (after approximately 500,000 cycles) for a given cyclic stress range or cyclic strain range. Figure 19 illustrates the agreement between prediction and experiment. The material is $[0_2/\pm45]_s$ boron/aluminum, with a fiber volume fraction of 0.45. The solid line represents the predicted laminate stiffness loss. Since the relation between cyclic stress range and cyclic strain range does not depend upon stress ratio, either can be used to predict secant modulus degradation. The secant modulus is shown to decrease with increasing stress or strain range. This particular laminate's secant modulus dropped over 40 percent without impending failure. The data were generated under either strain-controlled or load-controlled conditions.

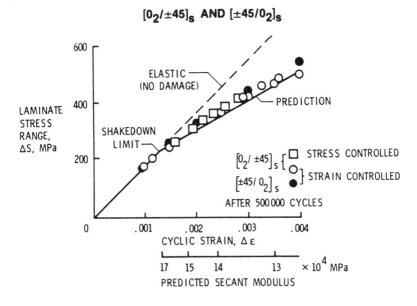

Figure 19

DEBONDING OF ADHESIVELY BONDED COMPOSITES UNDER FATIGUE LOADING

An experimental and analytical investigation based on fracture mechanics methodology was undertaken to study fatigue failure (such as the gradual growth of debonded area when loads are repeated) of adhesively bonded composite joints. A cracked lap-shear specimen was used to test the adhesives. This specimen simulates a realistic adhesive joint where a combination of shear and peel stresses are present. Different configurations of cracked lap-shear specimens are being investigated. The different configurations have different percentages of the strain energy release rate, G, associated with debond opening tension, G_I, and debond shearing, G_{II} (ref. 18). Specimens were made from T300/5208 graphite/epoxy and Kevlar/epoxy (ref. 19). The EC-3445 and FM-300 adhesives tested were cured at 121°C and 177°C, respectively. Specimens are being studied for the combination of both composites and adhesives. For example, figure 20 shows the correlation between the experimentally measured debond growth rate, da/dN, and the value of strain energy release rate, G_I, for the two adhesive systems measured using graphite/epoxy specimens. The FM-300 has a slower growth rate than EC-3445 for an equivalent G_I. The purpose of the program is to develop a model to relate strain energy release rate to debond behavior for any structural geometry. This would then in turn be used to design safe, efficient, bonded composite structures.

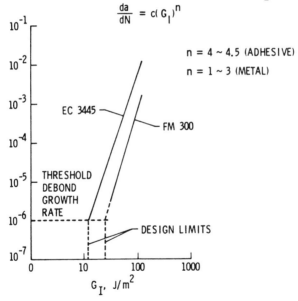

Figure 20

FUTURE WORK

As a result of these recent activities, work is planned in the following areas.

    o Develop analyses for buffer strips with a wide range of parameters.

    o Analyze damage mechanisms for bearing loaded holes.

    o Extend finite-element analysis of bolt clampup to account for transient environments.

    o Extend 3-D finite-element analyses of interlaminar stresses to account for thermal and hygroscopic effects.

    o Determine static and fatigue delamination failure criteria for a variety of mixed-mode loadings.

    o Predict delamination growth due to local instabilities.

    o Predict fatigue endurance limits for arbitrary composite laminates.

    o Predict fatigue endurance limits for adhesively bonded joints.

    o Evaluate second-generation composites for improved delamination and impact resistance.

REFERENCES

1. Poe, C. C., Jr.; and Sova, J. A.: Fracture Toughness of Boron/Aluminum Laminates With Various Proportions of 0° and ±45° Plies. NASA TP-1707, 1980.

2. Poe, C. C., Jr.: A Single Fracture Toughness Parameter for Fibrous Composite Laminates. NASA TM-81911, 1981.

3. Poe, C. C., Jr.; and Kennedy, J. M.: An Assessment of Buffer Strips for Improving Damage Tolerance of Composite Laminates. Journal of Composite Materials Supplement, vol. 14, no. 1, 1980, pp. 57-70.

4. Crews, J. H., Jr.: Bolt-Bearing Fatigue of a Graphite/Epoxy Laminate. Joining of Composite Materials, K. T. Kedward, ed., ASTM STP 749, American Society for Testing and Materials, 1981, pp. 131-144.

5. Shivakumar, K. N.; and Crews, J. H., Jr.: Bolt Clampup Relaxation in a Graphite/Epoxy Laminate. NASA TM-83268, 1982. (Also to appear in Long-Term Behavior of Composites, T. K. O'Brien, ed., American Society for Testing and Materials Special Technical Publication.)

6. Shivakumar, K. N.; and Crews, J. H., Jr.: An Equation for Bolt Clampup Relaxation in Transient Environments. NASA TM-84480, 1982. (Also accepted for publication in Composites Technology Review.)

7. Raju, I. S.; and Crews, J. H., Jr.: Three-Dimensional Analysis of $[0/90]_s$ and $[90/0]_s$ Laminates with a Central Circular Hole. NASA TM-83300, 1982. (Also accepted for publication in Composites Technology Review.)

8. Bostaph, G. M.; and Elber, W.: Static Indentation Tests on Composite Plates for Impact Susceptibility Evaluation. Paper presented at the U.S. Army Symposium on Solid Mechanics: Critical Mechanics Problems in Systems Design (Cape Cod, MA), Sept. 21-23, 1982.

9. O'Brien, T. K.: Characterization of Delamination Onset and Growth in a Composite Laminate. Damage in Composite Materials, K. L. Reifsnider, ed., ASTM STP 775, American Society for Testing and Materials, 1982, pp. 140-167. (Also NASA TM-81940, 1981.)

10. O'Brien, T. K.; Johnston, N. J.; Morris, D. H.; and Simonds, R. A.: A Simple Test for the Interlaminar Fracture Toughness of Composites. SAMPE Journal, vol. 18, no. 4, 1982, pp. 8-15.

11. Standard Tests for Toughened Resin Composites. NASA RP-1092, 1982.

12. Raju, I. S.; and Crews, J. H., Jr.: Interlaminar Stress Singularities at a Straight Free Edge in Composite Laminates. Journal of Computers and Structures, vol. 14, no. 1-2, 1981, pp. 21-28.

13. Whitcomb, J. D.: Finite Element Analysis of Instability-Related Delamination Growth. Journal of Composite Materials, vol. 15, 1981, pp. 403-426.

14. O'Brien, T. Kevin: Tension Fatigue Behavior of Quasi-Isotropic, Graphite/Epoxy Laminates. Fatigue and Creep of Composite Materials, H. Lilholt and R. Talreja, eds., Risø National Laboratory, Roskilde, Denmark, 1982, pp. 259-264.

15. Phillips, E. P.: Effects of Truncation of a Predominantly Compression Load Spectrum on the Life of a Notched Graphite/Epoxy Laminate. Fatigue of Fibrous Composite Materials, ASTM STP 723, American Society for Testing and Materials, 1981, pp. 197-212. (Also NASA TM-80114, 1979.)

16. Johnson, W. S.: Mechanisms of Fatigue Damage in Boron/Aluminum Composites. Damage in Composite Materials, K. L. Reifsnider, ed., ASTM STP 775, American Society for Testing and Materials, 1982, pp. 83-102. (Also NASA TM-81926, 1980.)

17. Johnson, W. S.: Modeling Stiffness Loss in Boron/Aluminum Below the Fatigue Limit. NASA TM-83294, 1982. (Also to appear in Long-Term Behavior of Composites, T. K. O'Brien, ed., American Society for Testing and Materials Special Technical Publication.)

18. Dattaguru, B.; Everett, R. A., Jr.; Whitcomb, J. D.; and Johnson, W. S.: Geometrically Nonlinear Analysis of Adhesively Bonded Joints. Paper presented at 23rd AIAA/ASME/ASCE/AHS Structures, Structural Dynamics, and Materials Conference (New Orleans, LA), May 10-12, 1982. (Also accepted for publication in Journal of Engineering Materials and Technology.)

19. Mall, S.; Johnson, W. S.; and Everett, R. A., Jr.: Cyclic Debonding of Bonded Composites. Paper presented at International Symposium on Adhesive Joints: Their Formation, Characteristics, and Testing (Kansas City, MO), September 13-16, 1982.

PROCESSING COMPOSITE MATERIALS

R.M. Baucom

NASA Langley Research Center
Hampton, VA

INTRODUCTION

Langley Research Center has an active role in the development of composite materials for aerospace applications. This activity was initiated over ten years ago and has included a variety of study efforts that have led to several production commitments to use composite materials in structural components. The overall composites program at Langley includes basic polymer studies, composites processing research, materials evaluation, analysis, and applications. In order to support the composite research activities, the current annual consumption of graphite, Kevlar and fiberglass reinforced polymer materials is approximately 1000, 400, and 800 pounds, respectively. An additional 50 pounds of fiber reinforced metal matrix composite materials are also consumed. Annually, approximately 300 structural laminates ranging in size up to 20 square feet and in thickness up to 1.2 inches are fabricated from these materials. Structural articles weighing up to several hundred pounds are also frequently fabricated. In addition to conventional vacuum molding capability, 12 heated platen presses, 4 research autoclaves, 2 thermoforming machines and a pultruder are available for composites fabrication. This paper will review composite processing methods and illustrate selected examples of components produced by Langley as well as components produced by our aerospace contractors.

COMPOSITE PROCESSING AND APPLICATIONS

Fiber reinforced polymer composite materials can be processed into structural articles by a vareity of different methods. The application of pressure or force to composite materials to shape and cure is accomplished by means of the application of vacuum, autoclave pressurization, trapped rubber expansion or hydraulic presses. Vacuum molding is a widely used method for the fabrication of large production, non-aerospace articles from general use composite materials such as fiberglass reinforced polyester and epoxy. Advanced composite materials such as Kevlar and graphite reinforced epoxies and polyimides are generally processed by autoclave molding, by trapped rubber expansion, or by hydraulic presses. To complete the processing cycle the articles fabricated by these methods are normally subjected to nondestructive inspection to verify the structural integrity of the finished part.

This presentation will feature the fabrication of several composite structural articles including DC-10 upper aft rudders, L-1011 vertical fins and composite biomedical appliances. Also, a discussion on innovative composite processing methods will be included.

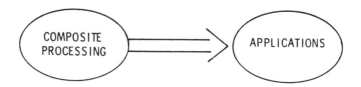

CONVENTIONAL

- VACUUM MOLDING
- AUTOCLAVE
- TRAPPED RUBBER
- PRESS MOLDING
- INSPECTION

- DC-10 UPPER AFT RUDDER
- L-1011 VERTICAL FIN
- BIOMEDICAL APPLIANCES

ADVANCED

- INTEGRATED LAMINATING CENTER
- HOT MELT FUSION

COMPOSITE MOLDING METHODS

The three most widely used methods for applying pressure to mold composites are thermal expansion of trapped rubber, autoclave pressurization, and hydraulic presses. The selection of the appropriate molding method depends primarily on the structural article size and maximum cure temperature and pressure. The trapped rubber or thermal expansion molding technique has the capability of developing very high pressures during composite cure and care must be exercised in the design of the tooling in order to maintain the selected pressures. This method is utilized in composite processing where uniform pressure application in complex shapes is difficult with other methods. Composite article size is limited only by oven size and tool mass with the thermal expansion molding technique. Autoclave processing of composite materials is widely used due to the precise control of pressure and temperature offered by the method.

Structural article size and maximum pressure are limited by the design of the autoclave pressure vessel. Hydraulic presses equipped with heated platens are generally utilized for applications which require high pressures and fast cycle times during composite molding. Hydraulic presses are widely used to fabricate small, intricate parts in matched metal tooling.

THERMAL EXPANSION

AUTOCLAVE

HYDRAULIC PRESS

COMPOSITE MOLDING BY THERMAL EXPANSION

Most rubber compounds possess the physical characteristic of expanding upon the application of heat. This characteristic can be used to apply uniform pressure to composite materials to form and cure a structural article. Typically, a block or plug of rubber is cast into the desired shape with appropriate allowances for expansion prior to contacting the mold cavity during heat up. The composite material is applied over the rubber plug and the assembly is placed in the mold cavity which serves as the pressure containment chamber during cure. As the temperature is increased the composite material is forced against the mold. Since the rubber expands uniformly, the composite material is subjected to near-hydrostatic pressure as the temperature approaches the cure temperature. After an appropriate dwell time at the cure temperature the assembly is cooled and the part is removed. The cast rubber mold is then removed from the part and the assembly is prepared for the next fabrication cycle. The cast rubber block can be used for several cycles without degrading the material or losing its pressurization capability which, in turn, reduces the tooling costs.

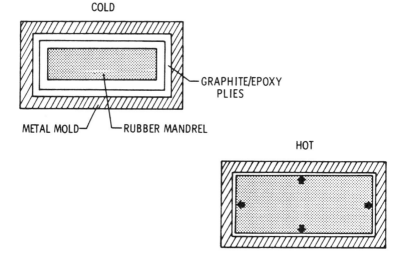

AUTOCLAVE MOLDING OF COMPOSITES

Autoclaves are utilized for processing a wide variety of composite materials into flat laminates and structural shapes. A typical autoclave process begins with assembling the part from the appropriate composite prepreg material. This assembly, commonly referred to as the layup, is placed on a metal caul plate that has been coated with spray release agent or a nonporous release film. The composite material is then covered with a porous release film and the required number of plies of breather material. A vacuum bag is then installed over the assembly and sealed around its periphery with gasket sealant material. A vacuum line is attached to the caul plate, vacuum is applied to the part, and the entire assembly is inserted into the autoclave. During heat-up to the composite cure temperature the viscosity of the resin in the prepreg material is lowered and the excess resin is drawn into the breather plies through the porous release fabric. Volatile products generated during cure are also removed in the same manner. After the temperature and autoclave pressure are held for the appropriate time to cure the prepreg, the vacuum bag assembly is removed from the autoclave, the vacuum bag is stripped away, and the fully cured part is removed. The materials utilized in this fabrication process are selected to withstand the temperature and pressure required for the particular composite material being processed. Autoclave molding is the most widely used process in the aerospace industry for manufacture of large graphite/epoxy and graphite/polyimide structural articles.

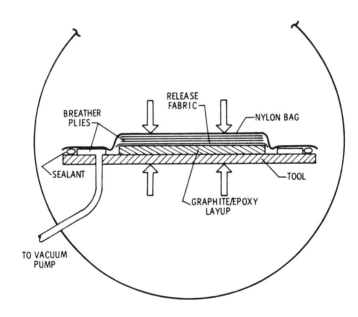

TYPICAL COMPOSITE CURE CYCLE

Polymer matrices for fiber reinforced composite materials generate reaction by-products during cure. In addition, the viscosity of the polymer matrix varies substantially during cure. In order to accommodate these physical phenomena during composite processing, the application of heat, vacuum, and pressure must be precisely controlled to avoid incomplete cures, voids, delaminations, and excess resin and fiber movement. A typical cure cycle for fabricating graphite/epoxy is shown in the figure. Vacuum is applied to the composite material and it is heated to an intermediate temperature of approximately 250-275°F. This condition is held for a period of time to allow excess solvent, water and reaction by-products to be removed. Prior to matrix gellation, the vacuum is removed and positive pressure is applied. The composite material is heated to the cure temperature, typically 350°F, and allowed to soak for 1 to 2 hours to effect a complete matrix cure. The part is then cooled to room temperature, the pressure is vented, and the part is removed. With minor variations to accommodate different composite material combinations, this profile is representative of most engineering composite material cure cycles.

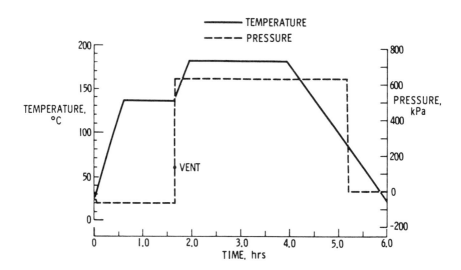

Selected Additional Subjects   403

ULTRASONIC INSPECTION OF COMPOSITES

After composite articles are fabricated it is necessary to establish the structural integrity of the finished part. An initial visual examination is performed to identify areas of gross disbonding, delamination, fiber misalignment, laminate cracking, warped areas, etc. This inspection is normally followed by nondestructive ultrasonic evaluation of the article to ensure that the article is free of internal defects. This method requires that the article be acoustically coupled to the ultrasound transducer. The most convenient way to establish positive acoustic coupling is to immerse the article in a water tank equipped with a traveling bridge for attachment of the transducer. As the transducer travels along the bridge, sound is directed through the water and into the article. When a discontinuity is detected in the composite, the decibel level of the sound transmitted back to the transducer is reduced. This data can be displayed on an oscilloscope in the form of signal amplitude changes or on a printer which highlights the defect area. Voids, delaminations, cracks and porosity absorb the sound transmitted by the transducer and can be readily identified by this inspection process. Transducer frequency, focal distances, and focal diameters are selected to accommodate article thickness, shape, and fiber and resin type.

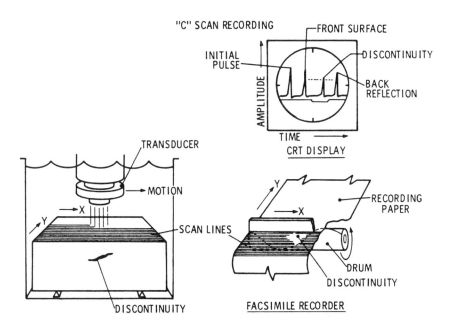

INSPECTION TECHNIQUES

The capability of ultrasonic inspection to verify the structural integrity composites has been demonstrated by machining, polishing, and visually inspecting the suspected defect areas. The ultrasonic C-scan image, shown in the upper left of the figure, indicates a structurally sound laminate. The small black dots on the C-scan represent the metal support pins used to elevate the laminate off the base of the water immersion tank to avoid undesirable artifacts in the C-scans. The photomicrograph of the laminate shown in the upper right of the figure verified the assessment. The ultrasonic C-scan in the lower left portion of the figure shows evidence of internal voids which absorbed the sound beam during inspection. This laminate was also sectioned, polished and visually inspected to verify the ultrasonic display of voids. Large discontinuities between laminate plies are evident in the defective areas indicated by ultrasonic inspection. The criteria for accept/reject of a composite article vary widely as a function of the data base generated for the particular composite system and the mission for the composite article. In particular, the criticality of void size, type, and location plays a major role in establishing the acceptable limits for anomalies discovered by ultrasonic inspection.

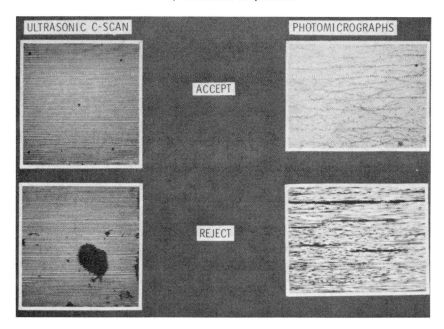

DC-10 UPPER AFT RUDDERS

One of the composite structural components developed under the NASA ACEE (Aircraft Energy Efficiency) program is the upper aft rudder for the DC-10 aircraft. The component is approximately 13 feet in length. The aluminum production design weighs 93 pounds whereas the composite rudder weighs only 61 pounds. The structural box was fabricated from graphite/epoxy material. The leading and trailing edges were fabricated from glass/epoxy material.

COMPOSITE DC-10 UPPER AFT RUDDER ADVANTAGES

The graphite/epoxy upper aft rudder for the DC-10 was fabricated using the trapped rubber thermal expansion technique. Principal advantages of using this fabrication procedure were: (1) the component was molded to net size and machining operations were minimized, (2) the complete assembly was cured in one cycle thereby saving costs and limiting exposure time of the composite to 350°F temperatures, and (3) the requirement for secondary bonding of subassemblies was eliminated. Composite rudders fabricated in this manner weigh approximately 30 percent less than the production aluminum rudders.

GRAPHITE/EPOXY COCURED COMPONENT

- MOLDED NET TO SIZE
- CURED IN ONE CYCLE
- NO SECONDARY BONDING
- SIGNIFICANT WEIGHT REDUCTION OVER METAL CONFUGURATION

Selected Additional Subjects 407

DC-10 UPPER AFT RUDDER MANUFACTURING SEQUENCE

Fabrication of the DC-10 graphite/epoxy upper aft rudder began with layup and pre-densification of the skins, front and rear spars and ribs from unidirectional tape and broadgoods. These parts were then loaded into the rubber molding tool along with the internal metal and rubber mandrels. Steel side plates were bolted into place and the assembly was placed in the oven. The assembly was heated to 350°F and held for 2 hours and 15 minutes to fully cure the graphite/epoxy. The finished rudder was removed from the tool.

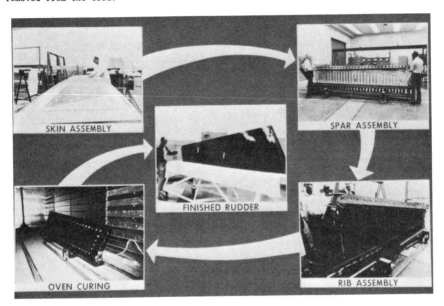

THERMAL EXPANSION MOLDING TECHNIQUE

Thermal expansion molding technique (trapped rubber processing) is predicated on the fact that silicone rubber, when confined within a vessel and subjected to heat, thermally expands and generates internal pressure. As in the case of the upper aft rudder of the DC-10, graphite/epoxy details are laid up, densified on wooden form blocks, trimmed and prepared for freezer storage. The formed details are then loaded into the rudder tool, the rubber mandrel is properly positioned, and the tool vessel is closed. The loaded tool is then placed in an oven and cured for 2 hours at 350°F after initial heat-up. Advantages of the process include elimination of the autoclave and vacuum bagging operations.

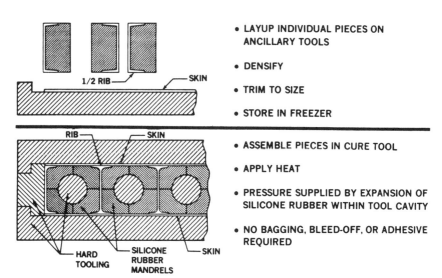

- LAYUP INDIVIDUAL PIECES ON ANCILLARY TOOLS
- DENSIFY
- TRIM TO SIZE
- STORE IN FREEZER

- ASSEMBLE PIECES IN CURE TOOL
- APPLY HEAT
- PRESSURE SUPPLIED BY EXPANSION OF SILICONE RUBBER WITHIN TOOL CAVITY
- NO BAGGING, BLEED-OFF, OR ADHESIVE REQUIRED

L-1011 ADVANCED COMPOSITE VERTICAL FIN

Two structural components on the L-1011 airplane were selected for fabrication from graphite/epoxy composite as part of the NASA ACEE program. One component, the vertical fin, is a primary load-carrying structural element. The objectives of the Advanced Composite Vertical Fin (ACVF) research were to develop low-cost manufacturing processes for large composite structural aircraft articles and to verify the structural integrity and durability of the article.

L-1011 ACVF STRUCTURAL CONFIGURATION

The ACVF is 25 feet long, 9 feet wide at the root, and weighs 622 pounds. It is comprised of 10 ribs, front and rear spars, and hat-stiffened cover skins. Unidirectional and woven graphite/epoxy prepreg is utilized in the construction of the ACVF subassemblies.

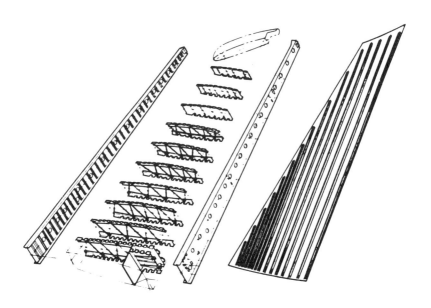

Selected Additional Subjects 411

ACVF SKIN LAY-UP SCHEMATIC

The skin of the L-1011 composite fin (ACVF) is comprised of a ply buildup which is 34 plies at the root end (to match the existing L-1011 metallic mating structure) and tapers progressively to 16, 14, and 10 ply areas as shown. The buildup consists of five plies over the complete fin area, oriented ±45°, 0°, ∓45°. This is followed by 10 partial plies, principally 0° orientation, building up toward the root end, and a core of four ±45° plies at the midpoint in the symmetrical layup. Additional partial plies are included at the tip and along the front and rear spar attach areas to provide a reinforced bearing pad for fastening.

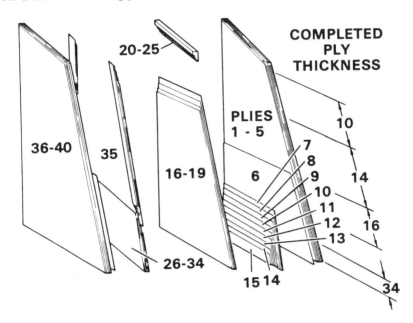

ACVF HAT/SKIN VACUUM BAG ASSEMBLY

The various types of materials which are used in the cure of a typical hat-stiffened bay of the ACVF cover assembly are shown in the figure. A barrier film is used next to the tool surface and also under the hat cauls to prevent resin adhesion to the tooling components. A nylon peel ply is placed immediately adjacent to the graphite layup, both on the tool side of the skin and over the hats and between hat flanges. Bleeder material is located between hat flanges and around the periphery of the part. Breather plies cover the completed stack to assure a continuous vacuum path to the vacuum ports located around the tool base. A vacuum bag covers the complete assembly and is sealed at the edges of the tool. The inflatable bladder penetrates the vacuum bag to admit autoclave pressure to the bladder interior.

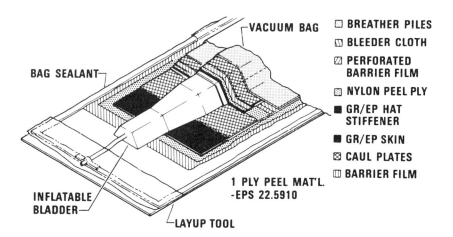

Selected Additional Subjects 413

ACVF COVER FABRICATION

After the ACVF components are laid up, densified, and cured they are removed from the autoclave. The vacuum bag assembly for the fin cover shown in the figure has been removed and the inflatable rubber mandrels utilized for pressurization of the hat stiffened elements during cure are being removed.

414 Tough Composite Materials

COMPLETE GRAPHITE/EPOXY ACVF COVER

The graphite/epoxy ACVF cover skin with integral cocured hat stiffener elements is shown being removed from the autoclave cure tool. After removal from the tool, the covers are subjected to a complete ultrasonic scan inspection to ensure that the part is free of voids, delaminations, or other anomalies that may affect structural integrity. The skins are then machined for mechanical attachment to the rib and spar assemblies.

Selected Additional Subjects   415

AUTOMATED INTEGRATED MANUFACTURING SYSTEM

During the 1970's, as composites became more generally accepted, the need for mechanized equipment to fabricate severely contoured, integrally stiffened, complex structures for aircraft was identified. This led to the development and implementation of a number of automated composites processing centers in the aircraft industry. The Automated Integrated Manufacturing System (AIMS) shown in the figure was developed by the Grumman Aerospace Corporation under Air Force sponsorship. This system has the capability to automatically dispense and laser trim composite tape and broadgoods. The prepreg details can then be automatically transferred to a contour ply handler or translaminar stitching machine for final layup and assembly into structural articles.

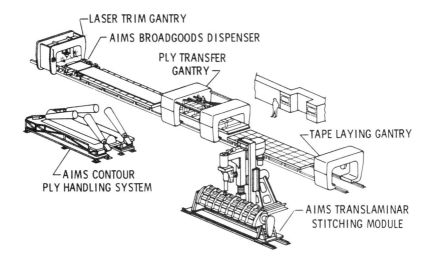

INTEGRATED LAMINATING CENTER

The heart of the Grumman AIMS is the Integrated Laminating Center which is comprised of three traveling gantrys and two separate layup racks. The composite tape or broadgoods are dispersed onto the appropriate rack and trimmed by the laser trim gantry. The composite details can then be removed by the ply transfer gantry and placed on the appropriate layup tool.

HOT MELT FUSION COMPOSITES

The state of the art for fabrication of fiber reinforced thermoplastic composite materials requires the impregnation of the reinforcement fiber with thermoplastic polymers that have been placed in solution with appropriate solvent systems. The matrix polymer materials that lend themselves to being placed in solution for impregnation naturally have limited resistance to exposure to various solvents after fabrication of a structural article. One obvious solution to this problem is to utilize a solvent-resistant polymer matrix material for impregnation of the composite. Since these matrix materials cannot be readily placed in solution, the reinforcing fibers must be impregnated by some form of hot-melt fusion of the matrix polymer. Performing this function by coating the fibers with films has met with limited success due to the inability to penetrate fiber bundles with the high-viscosity, solvent-resistant matrix materials. The technique of blending reinforcing fibers with nearly equal dimater thermoplastic matrix fibers is under investigation in the Materials Division at Langley Research Center. Preliminary results with laboratory-scale graphite/PBT thermoplastic composites fabricated in this manner indicate that very good wetting of the reinforcing fiber with the matrix fiber takes place during application of heat and pressure to form the composite. Work is under way to scale this technique up to fabricate broadgoods with graphite/thermoplastic PBT tows and subsequently manufacture structural articles from the graphite/PBT cloth.

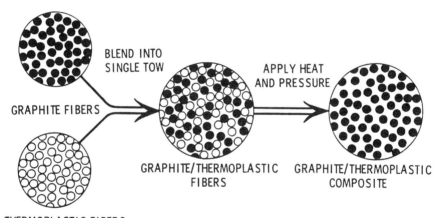

KEVLAR/EPOXY FACEGUARD

One physical impairment that frequently occurs as a result of cerebral palsy is the onset of seizures ranging from very mild to violent. Uncontrolled falls frequently result from these seizures which, without proper protection, can cause serious injury to the upper body and facial area of a person. One of the standard protection devices for uncontrolled falls is a helmet very similar to a football linemans' helmet. Due to the very heavy weight of this type of helmet, patients often refuse to wear them, which in turn eliminates the protection afforded by the helmet. A prototype Kevlar/epoxy protective face protection mask was fabricated at Langley for the boy shown in the figure. The mold for the mask was made from plaster and four plies of epoxy-impregnated Kevlar cloth were applied to the mold. After curing the mask it was removed from the plastic mold and the mouth and eye openings were cut. The resultant weight of this composite mask was 4 ounces compared to approximately 2 pounds for the football lineman's helmet. The patient in the figure has freely worn the Kevlar/epoxy mask for more than a year and has sustained no injuries in this time period. Prior uncontrolled falls without protection had resulted in several serious injuries to this patient including multiple jaw and nose breaks and loss of several permanent teeth.

COMPOSITE WHEELCHAIR PROJECT

NASA Langley Research Center and the University of Virginia Rehabilitation Engineering Center are working together to design, fabricate, and evaluate a durable, lightweight composite wheelchair for general use. The graphite/epoxy wheelchair shown in the figure was the forerunner for the current NASA/UVA effort. All of the major structural components for the wheelchair in the figure were fabricated from graphite/epoxy to develop confidence in the fabricability of wheelchair elements from composites. This wheelchair was designed to transport invalid passengers aboard aircraft. The current general-use wheelchair under development will employ a variety of composite materials in its construction. The side, seat, and foot rest elements will be fabricated from a composite system composed of graphite/epoxy and Kevlar/epoxy hybrid skins bonded to each side of a special high strength inert polyimide foam core. This structurally efficient composite system has been subjected to a variety of bending and flexural tests to identify the optimum combination of elements. The target weight for the composite wheelchair is 25 pounds, whereas conventional general-purpose metal wheelchairs weigh over 50 pounds. Three prototype composite wheelchairs are scheduled to be completed by March 1983 for structural and clinical evaluation.

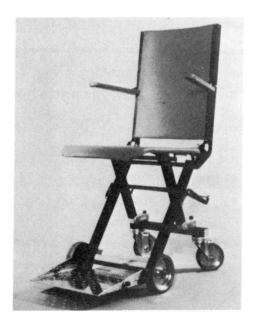

SUMMARY

Composite materials processing at Langley Research Center encompasses a wide variety of concepts and applications ranging from the construction of exotic models for research to the computer-assisted identification of critical molecular changes in polymer systems during cure. Along with the research support functions performed by highly specialized technicians, research is continuing in composite materials processing to identify and provide the concepts necessary to fabricate low-cost, reliable, and efficient structures. Facilities are continually being updated to provide support for this research. New composite systems formulated in-house are assessed on a continuing basis to establish their utility for a variety of commercial and aerospace applications.

CURRENT TECHNOLOGY

- ESTABLISHED FABRICATION PROCEDURES
- FLIGHT QUALITY COMPONENTS PRODUCED

FUTURE EMPHASIS

- AUTOMATION OF COMPOSITE PROCESSING
- PROCESSING TECHNIQUES FOR ADVANCED RESINS

SPINOFF

- BIOMEDICAL
- SPORTS
- TRANSPORTATION

OPPORTUNITIES FOR COMPOSITES IN COMMERCIAL TRANSPORT STRUCTURES

Herman L. Bohon

NASA Langley Research Center
Hampton, VA

ACEE COMPOSITES PROGRAM-

INTRODUCTION

In recent years graphite/epoxy material has found widespread application in military aircraft and is now finding application in commercial transports. Because of special features of this material, such as high strength-to-density ratio, good formability and laminate tailoring, the next generation of military and commercial aircraft manufactured with composites could be significantly more efficient than current aircraft. Studies have shown that composites can reduce the structural weight of transport aircraft by as much as 25 percent over current aluminium structures with a corresponding reduction in fuel consumption of 12 to 15 percent. The NASA Aircraft Energy Efficiency (ACEE) composites program was established to foster the application of composite material in the next generation of aircraft. The primary objective of ACEE is to develop the essential technologies in cooperation with the commercial transport manufacturers to permit the efficient utilization of composites in airframe structure of future transport aircraft.

Specifically the ACEE composites program is to provide each of the commercial transport manufacturers both the technology and confidence required for a commitment to composite structures production. This means not only know-how for predictable designs and low-cost fabrication, but enough test and actual manufacturing experience to accurately predict durability for product warranty purposes and costs for product pricing, and to assure safety for certification by the FAA and maintainability for acceptance by the airlines.

OBJECTIVE

PROVIDE THE **TECHNOLOGY** AND **CONFIDENCE** SO THAT COMMERCIAL TRANSPORT MANUFACTURERS CAN COMMIT TO PRODUCTION OF COMPOSITES IN THEIR FUTURE AIRCRAFT:

SECONDARY STRUCTURE - 1980 TO 1985

PRIMARY STRUCTURE - 1985 - 1990

TECHNOLOGY

- DESIGN CRITERIA, METHODS AND DATA
- QUALIFIED DESIGN CONCEPTS
- COST COMPETITIVE MANUFACTURING PROCESSES

CONFIDENCE

- DURABILITY / WARRANTY
- QUANTITY COST VERIFICATION
- FAA CERTIFICATION
- AIRLINE ACCEPTANCE

COMPONENT DEVELOPMENT TOWARD A PRODUCTION COMMITMENT

To support these objectives the manufacturers are developing composite versions of structural components on existing aircraft with NASA paying 90 percent of the cost. Development involves testing of various material options before selecting one and then extensive testing to develop an adequate data base of material strength and stiffness properties. Design options are narrowed through analysis and a varied spectrum of development tests on small and large subcomponents. In parallel with this, a suitable production process including economical ply preparation and cure at high temperature and pressure is evolved, tools are designed and fabricated, and full-scale components are then manufactured for ground qualification tests, flight tests, and airline service. The various tests include many that are required by the FAA for flight certification, which must precede airline service. Inspection and repair methods to insure adequate maintenance in service are also developed.

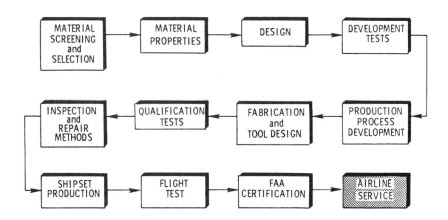

ACEE COMPOSITE SECONDARY STRUCTURES

The commercial transport manufacturers were challenged to redesign selected secondary and medium primary components on existing aircraft with composite material. Secondary components include the Boeing 727 elevators, the Lockheed L-1011 ailerons and the Douglas DC-10 upper aft rudder. Such components are called "secondary" structures because they do not carry primary flight loads and are not critical to flight safety.

BOEING 727 COMPOSITE ELEVATOR

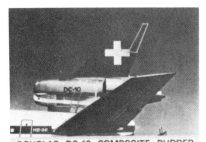

DOUGLAS DC-10 COMPOSITE RUDDER

LOCKHEED L-1011 COMPOSITE AILERON

DC-10 GRAPHITE/EPOXY UPPER AFT RUDDER

The DC-10 composite rudder program is now complete and one of the rudders is shown here. The rudder is 3 feet at the root chord by 13 feet in length and weighs only 67 pounds, corresponding to a 26-percent weight savings over the metal design. Douglas has fabricated 20 rudders under this program and to date 13 have been placed in flight service with several domestic and foreign commercial airlines.

727 GRAPHITE/EPOXY ELEVATOR

In the Boeing 727 elevator program five shipsets of composite elevators have been fabricated and ground and flight tested. Four shipsets are shown here. This structure has been certificated by the FAA and all five shipsets have been placed into airline service with United Airlines. The elevators are 3-1/2 feet at the maximum chord by 17 feet in span and weigh 98 pounds. Weight reduction over the metal design is about 25 percent. The design and construction of these elevators provided the confidence and experience to place generically similar composite designs of secondary components on Boeing's new 767 and 757 aircraft.

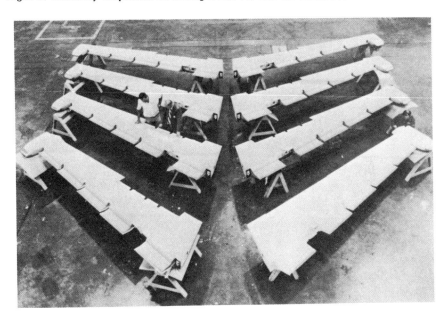

L-1011 GRAPHITE/EPOXY AILERON

The Lockheed L-1011 aileron program is also complete. Five shipsets of ailerons were successfully fabricated, ground and flight test qualified and certificated by FAA. One shipset is in service on the Lockheed Company airplane while the other four are in flight service on commercial aircraft, two with TWA and two with Delta. The ailerons are 4 feet by 8 feet and weigh 107 pounds, representing a 24-percent weight savings over the metal design.

INSTALLED ON L-1011 WING

ACEE COMPOSITE MEDIUM PRIMARY STRUCTURES

The commercial aircraft manufacturers are also developing composite versions of empennage primary structures on their existing aircraft. The empennage components offer a significant increase in challenge for composites application compared to the secondary structures. In particular, physical size, design requirements, load interaction, manufacturing and tooling each present formidable technology development tasks beyond those required for secondary structure. Components selected for redesign are the horizontal stabilizers on the Boeing 737 and the vertical stabilizers of the Douglas DC-10 and the Lockheed L-1011. These programs are still in progress.

BOEING 737 COMPOSITE HORIZONTAL STABILIZER

DOUGLAS DC-10 COMPOSITE VERTICAL STABILIZER

LOCKHEED L-1011 COMPOSITE VERTICAL FIN

DC-10 GRAPHITE/EPOXY VERTICAL STABILIZER

All subcomponents (covers, spars, and ribs) of three full-scale composite vertical stabilizers for the DC-10 have been successfully fabricated and one stabilizer has been assembled for detailed ground qualification testing. The stabilizer is 7 feet at the root chord and 23 feet in span and weighs 779 pounds, representing a 23-percent weight savings over the metal design. The ground test program, which began in December 1981, was interrupted by premature failure initiating at the rear spar. Cause of failure has been identified and design modifications incorporated in the other two units. The second unit is being assembled to complete ground qualification testing, and the third unit will be flight tested to complete requirements for FAA certification. The third unit will then be placed in flight service with a commercial airline.

737 GRAPHITE/EPOXY HORIZONTAL STABILIZER

The 737 composite horizontal stabilizer is the smallest of the three empennage components and measures 4 feet at the root chord by 17 feet in span. The composite stabilizer weighs 204 pounds, which is a 22-percent weight reduction from the metal design. All ground and flight testing was completed successfully and FAA certification was granted in August 1982. This marks the first FAA certification of an empennage component for commercial transport aircraft. Boeing has fabricated and assembled five shipsets which are expected to be placed in flight service in the near future.

L-1011 GRAPHITE/EPOXY VERTICAL FIN

The L-1011 composite vertical fin is the largest in planform of the empennage components with a root chord of 9 feet and a span of 25 feet. The composite fin weighs 622 pounds, which is 28 percent lighter than the metal design. Two complete L-1011 composite fins have been assembled and one of these failed during ground qualification test just below design ultimate load, revealing a minor but significant design flaw. (Details of the failure will be discussed later.) The other fin was strengthened and successfully completed ground tests in June 1982 to qualify the modified design. Flight testing of the L-1011 composite fin is not planned.

ACEE COMPOSITE COMPONENT STATUS

Key features of the three secondary and empennage primary components are summarized in this table. Each component was manufactured in a production environment as a direct replacement of existing metal components. Weight savings from 22 to 28 percent were obtained even though these composite designs were driven to some degree by the existing metal design requirements. All components placed in flight service are tracked by the airlines and the manufacturer and are inspected periodically to insure safety of flight.

COMPONENT	SIZE ROOT X SPAN FT.	METAL DESIGN WEIGHT (LBS.)	COMPOSITE DESIGN WEIGHT (LBS.)	WEIGHT REDUCTION	NUMBER OF PRODUCTION UNITS	FAA CERTIFICATION	REMARKS
SECONDARY STRUCTURES							
DC-10 RUDDER	3.2 x 13.2	91	67	26.4%	20	MAY 1976	13 UNITS IN FLIGHT SERVICE
727 ELEVATOR	3.4 x 17.4	130	98	24.6%	11	JAN. 1980	5 SHIPSETS (10 UNITS) IN FLIGHT SERVICE
L-1011 AILERON	4.2 x 7.7	140	107	23.6%	12	SEPT. 1981	4 SHIPSETS (8 UNITS) IN FLIGHT SERVICE
MEDIUM PRIMARY STRUCTURES							
DC-10 V. STABILIZER	6.8 x 22.8	1005	779	22.6%	5	SEPT 1984	FLIGHT C/O AUG 1984
737 H. STABILIZER	4.3 x 16.7	262	204	22.1%	11	AUG. 1982	FLIGHT C/O COMPLETE
L-1011 V. FIN	8.9 x 25	858	622	28.4%	2	No	NO FLIGHT TEST PLANS

CHARACTERISTICS OF GRAPHITE/EPOXY MATERIAL

The remainder of this paper will highlight lessons learned in the ACEE composites program first by looking at attributes of the graphite/epoxy composite system and then by examining areas requiring advanced technology development.

Throughout the program the manufacturers required all fabrication and assembly of full-scale parts to be carried out in their production shops in order to obtain reliable manufacturing cost data as well as to gain valuable experience with production personnel. Hands-on experience was essential, not only in laying up and curing the laminate material with consistency and repeatability, but in developing test procedures to properly validate structural performance. This figure lists several positive characteristics of the composite system in manufacturing and performance areas that became evident during the full scale production phase of the program. The next few figures are used to illustrate these characteristics.

MANUFACTURING

- **TOLERANT RESIN CURE CYCLE**
- **UNIFORMITY OF HEAT DURING CURE OF PART IS ESSENTIAL**
- **PART SIZE LIMITED ONLY BY FACILITIES**
- **INNOVATIVE TOOLING PERMITS FABRICATION OF COMPLEX PARTS**

PERFORMANCE

- **FAILURE LOADS ARE PREDICTABLE**
- **FABRICATED PARTS ARE UNAFFECTED BY ENVIRONMENT**

TYPICAL CURE CYCLES FOR GRAPHITE/EPOXY

The epoxy resin system used in the ACEE composites program is cured at 350°F and at high pressure (80 to 100 psi depending on the part). The structural performance of the part is found to be tolerant of the heat-up rate, a feature which offers considerable flexibility in utilization of facilities. A typical cure cycle matrix is shown in this figure. The three curves outline the fastest, average, and slowest heat rises used in an empennage program. The left and center curves show a heat rise to 250°F under vacuum, a dwell period, and pressure application. The right curve with a much slower heat rise has pressure applied at 225°F without a dwell period. The heat sink capacity of the tool generally determines the heat-up rate and care must be taken to assure uniformity of heat of the part during the cure cycle to avoid irregular resin flow.

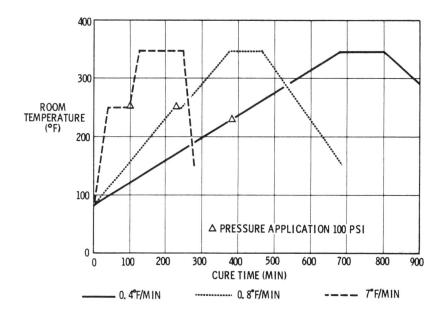

LARGE COMPLEX STRUCTURES MANUFACTURED

The largest single composite part in the empennage program is the cover panel of the L-1011 vertical fin. This structurally efficient hat-stiffened panel, which measures 8.9 feet at the root chord and 25 feet in span, was integrally cured in an autoclave and has no mechanical fasteners. The ease with which this part was cured suggests that the only limitation to part size may be the physical dimensions of the autoclave. The differential growth at cure temperature between the composite part and the hard tool (in this case steel) must, of course, be accounted for to assure dimensional control of critical elements and for tool release during cool-down following cure.

LIGHTWEIGHT STRUCTURAL COMPONENTS SIMPLIFY FINAL ASSEMBLY

One of the important features of the graphite/epoxy system is the unique opportunity to tailor stiffness properties and, with innovative tooling, to integrally mold efficient structural shapes that virtually eliminate mechanical fasteners. This figure illustrates one such example. The I-shaped spar configuration shown is 25 feet in length and is integrally (one step) cured, including the angle stiffeners on the shear web. Although some metallic parts are retained in final assembly, this single composite spar replaces 35 metal parts and 2,286 fasteners that were part of the all-metal design.

STRUCTURAL PERFORMANCE DEMONSTRATED

Following successful development of fabrication procedures for complex parts it is essential to validate the structural performance of cured parts. One program conducted by Lockheed to assess manufacturing tolerances on repeatability of performance is shown on the next three figures. Ten full-scale segments each of the L-1011 composite fin spar and cover were tested to destruction to obtain static strength characteristics. Each of the segments had measurable but acceptable manufacturing flaws such as thickness variations and small areas of porosity. The other twelve segments of each configuration are currently undergoing an accelerated 20 years of lifetime testing simulating flight cyclic loads, moisture, and temperature environments. Results from the static test program are shown in the next figure.

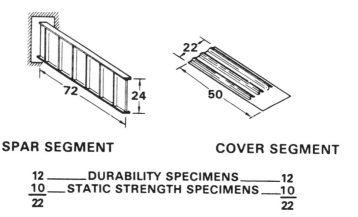

SPAR SEGMENT   COVER SEGMENT

$$\begin{array}{r} 12 \\ \underline{10} \\ 22 \end{array} \text{—DURABILITY SPECIMENS—} \begin{array}{l} 12 \\ \underline{10} \\ 22 \end{array}$$
STATIC STRENGTH SPECIMENS

SPAR STATIC STRENGTH TESTS FOR MANUFACTURING VARIANCES

The results from the static strength tests of ten duplicate spar segments are shown in this figure. The spars were loaded in bending to produce a distribution of strain in the critical region equal to that of the full-size spar under flight loads. The load at failure is plotted as a percent of design ultimate load (DUL), which is 1.5 times the maximum aerodynamic load expected in flight. The test average is 135 percent DUL and about 15 percent higher than the predicted value based on conservative material properties from coupon tests. More importantly, all test values exceeded the predicted value, suggesting that the allowable manufacturing flaws have little effect on static strength.

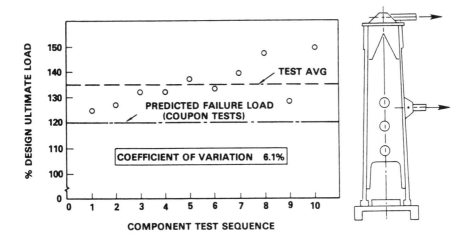

COMPARISON OF STATIC STRENGTH FOR
STRUCTURAL MATERIALS

The coefficients of variation of the ten cover and spar segments were 3.3 and 6.1, respectively. These values, obtained from tests of complex structural components, are consistent with values obtained from composite coupons and other structural materials and attest to the predictability and repeatability of performance of composite parts manufactured in a production surrounding.

MATERIAL	COMPONENT	SPEC. NO.	LOADING	COEFFICIENT OF VARIATION PERCENT
Graphite-Epoxy	PRVT-Cover	10	Compression	3.3
Graphite-Epoxy	PRVT-Spar	10	Bending	6.1
Graphite-Epoxy	Spoiler	15	Bending	6.6
Graphite-Epoxy	Laminate Coupons	411	Tension	5.7
Graphite-Epoxy	Laminate Coupons	411	Ten-Modulus	4.0
Graphite-Epoxy	Laminate Coupons	290	Compression	9.0
Graphite-Epoxy	Laminate Coupons	290	Compr-Modulus	5.2
Wood	Mosquito Wings	5	Bending	10.3
Wood	Plywood Shear Wall	27	Shear	9.7
Concrete	Test Cylinders	216	Compression	10.6
Aluminum	7049-T73 Die Forging	384	Tension	3.2
Aluminum	A357-T6 Casting	804	Tension	5.5
Titanium	TI-5AL-2.5SN Sheet	565	Tension	3.9
Steel	Structural Steel	3982	Tension	7.1
Steel	17-7PH Sheet	88	Tension	5.1

INFLUENCE OF ENVIRONMENTAL PARAMETERS ON STRUCTURAL PERFORMANCE

Another important issue that required full-scale components to evaluate was the influence of flight environmental parameters (moisture and temperature) on the performance of the structure under flight loads. A key activity designed to assess the influence of variations of flight temperature and moisture on strain distribution in a three-dimensional structure was carried out by Douglas on a stub box component shown in this figure. This component is the lower 45 percent of the DC-10 composite vertical stabilizer. The entire stub box, including the metallic leading edge, was mounted in an environmental chamber and subjected to a series of tests to determine, among other things, if the range of temperature and moisture to be expected in routine flight would affect the distribution of strain. The article was tested to limit load in bending, torsion, and shear at ambient conditions (72°F and no moisture (dry)), and then in bending at 0°F. This was followed by moisture conditioning at 170°F and 95 percent humidity for 14 days to achieve about 1 percent by weight moisture in the laminated structure. The component was then cyclic loaded for one lifetime at 0°F/wet and again tested to limit load in bending at 0°F and 130°F. Following a second lifetime of cyclic loading and another limit load bending test at 0°F/wet the stub box was then loaded in bending to failure. Measured strains for the different test conditions are shown on the next figure.

INFLUENCE OF MOISTURE AND TEMPERATURE ON STRAIN

The stub box strains shown are taken from the same strain gage at a cutout in the web of the rear spar. Data are plotted as a function of limit load for the four environmental conditions - ambient/dry, cold/dry, hot/wet, and cold/wet. The strain variation between the test conditions is small and suggests a negligible effect of moisture and temperature on strains in the three-dimensional structure.

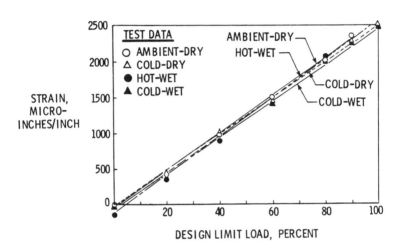

REAR SPAR CUTOUT STRAINS

KEY AREAS FOR TECHNOLOGY ADVANCEMENT

With the ACEE composites program now near completion it is interesting to examine the state of technology, especially in light of requirements of the future. While the current material system is clearly adequate for application to lightly loaded (secondary and empennage) structure, there are identifiable areas where technology advancement could significantly enhance the utilization of composites in primary wing and fuselage components. Three areas for technology advancement of particular importance to primary structure application are listed on this figure. Of these, secondary loads are probably the least understood. It is agreed among designers that there is a general state of uncertainty with composites as to the source, magnitude, and effects of secondary loads. Yet secondary loads are virtually impossible to eliminate from a complex built-up structure. While these loads are properly ignored in metallic structures, the sensitivity of current composite materials to interlamina forces can lead to serious weaknesses being designed into a composite structure. Such loads may be produced by eccentricities, irregular shapes, stiffness changes, and discontinuities, and their effects are magnified by the nonyielding nature of composites, which precludes load redistribution due to plasticity effects. The influence of secondary loads was vividly illustrated during ground tests of the L-1011 vertical fin, an event discussed on the two subsequent figures.

- **IMPROVED UNDERSTANDING OF SOURCE, MAGNITUDE, AND EFFECTS OF SECONDARY LOADS**

- **ADVANCED MANUFACTURING PROCESSES**

- **TOUGHER RESIN TO IMPROVE DAMAGE TOLERANCE AND DURABILITY**

EFFECTS OF SECONDARY LOADS IN COMPOSITES

The L-1011 composite vertical fin failed during ground verification tests at 98 percent of design ultimate load (DUL) - 28 percent less than the predicted failure load. The failure caused separation of the cover and front spar, as shown in this figure, along the entire length of the component. After a careful investigation, the cause of failure was determined to be due to secondary loads caused by local buckling of the cover at the cover/spar interface. While local buckling beyond limit load was allowed in the design, the influence of these loads on the integrity of the structure was not expected. The interlamina tension forces caused delamination of the spar cap and ultimately separation along the line of the fasteners. Results from post-failure tests to assess the influence of such loads are shown on the next figure.

TYPICAL SPAR CAP DAMAGE IN PRIMARY FAILURE ZONE

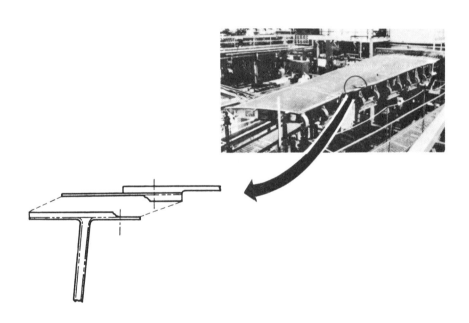

INFLUENCE OF LOAD CYCLING ON INTERLAMINA STRENGTH

Tests were conducted to determine the strength of the designs of the spar/cover intersection for secondary loads causing interlamina tension and transverse tension. The first test was on virgin material that had no prior loading and results were reasonable and indicated adequate margin. The estimated maximum interlamina tension load expected in flight was 68 pounds per fastener. The second test, on an undamage segment of the failed spar, showed large reductions in strength. This apparent influence of load cycling was verified by a third test on specimens subjected to load cycles similar to those of the ground test article prior to failure. This degradation in strength is the result of a design weakness in the spar cap, which can be avoided in future designs when criteria for secondary loads become a routine part of design practice.

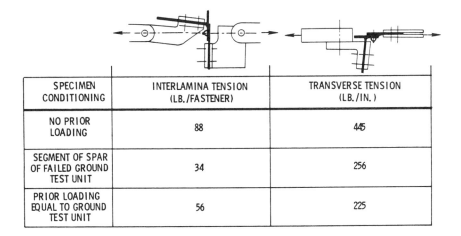

SPECIMEN CONDITIONING	INTERLAMINA TENSION (LB./FASTENER)	TRANSVERSE TENSION (LB./IN.)
NO PRIOR LOADING	88	445
SEGMENT OF SPAR OF FAILED GROUND TEST UNIT	34	256
PRIOR LOADING EQUAL TO GROUND TEST UNIT	56	225

INFLUENCE OF DAMAGE ON ALLOWABLE STRAIN

Another area in which technology advancement should pay off is the improvement of strain capability of the cured laminate. Limiting strain of current graphite/epoxy systems is determined primarily by sensitivity to impact damage. Typical results on this figure show the failure threshold of a compression-loaded graphite/epoxy plate as a function of impact energy. Designs of secondary and empennage structures have been governed primarily by stiffness requirements with design ultimate strains of about 0.003 μin./in.; and indeed, current resin systems can meet the strain requirements without a weight penalty. However, for large primary wing and fuselage structures, which are designed by strength requirements, higher strain capability is required if maximum utilization of composites is to be achieved. This is illustrated on the next figure.

EFFECT OF PROJECTILE IMPACT ON COMPRESSION STRENGTH

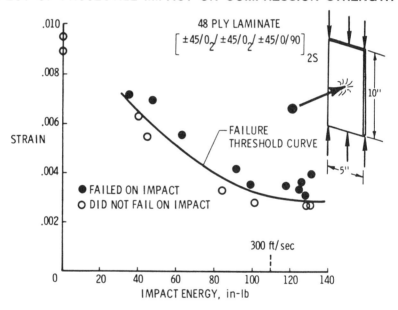

WEIGHT EFFECIENCY OF PRIMARY STRUCTURE DEPENDENT ON ALLOWABLE STRAIN

Wing structure of large transport aircraft carry considerably higher unit loads than other parts of the vehicle and element designs require thick laminates that are inherently stiff; consequently, strength rather than stiffness is the primary design driver. Recent wing surface design studies for commercial transports indicate that significant weight savings are permissible if design strains are not limited by impact requirements. This figure shows three structural configurations for wing surface panel designs; the curves on the figure show weight savings as a function of design strain for an optimized composite "blade" stiffened design over an aluminium design for the upper and lower surfaces. As can be seen a significant increase in weight savings for wing surface panels is possible if design strain can be increased from 0.004 to 0.006 μin./in. NASA is actively pursuing research in constituent relationships within the matrix and fiber/matrix interaction to achieve the higher strain capability.

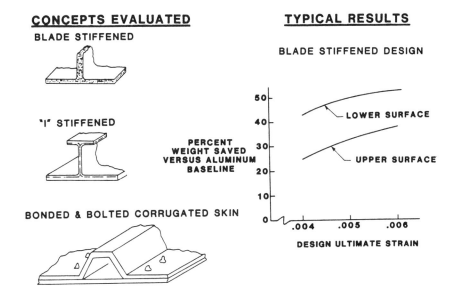

ACEE COMPOSITES PROGRAM CONCLUSIONS

The composite element of the NASA Aircraft Energy Efficiency (ACEE) program was initiated for the development of technology within the commercial airframe industry to foster the application of composites in future commercial aircraft. The program, which has been ongoing for nearly 6 years, has demonstrated the high potential for composites in commercial aircraft through the design and development of secondary and empennage primary components for existing aircraft. The composite material system used has been shown to be adaptable to variations in manufacturing processes, and it is readily formable into complex and highly efficient structural shapes with relatively good repeatability in performance. Large complex box structures have been fabricated, assembled, and ground tested and have provided a focus for technology advancements to further improve performance and assure flight safety.

The composites program has been very effective in developing confidence and experience within commercial airframe companies as engineering and manufacturing personnel have accepted and met challenges to develop and demonstrate weight and cost effective composite components. This level of confidence is strengthened by the generally widespread application of secondary composite components in flight service stemming from the ACEE composites program or prior NASA programs. This flight service experience is providing much needed airline participation and should pave the way for more committed involvement in composites in empennage primary and large primary structures in the next generation of aircraft.

- **LARGE STRUCTURAL COMPONENTS HAVE BEEN SUCCESSFULLY DESIGNED, FABRICATED, AND VERIFICATION TESTED**

- **WEIGHT SAVINGS UP TO 27 PERCENT DEMONSTRATED**

- **DESIGN AND MANUFACTURING CONFIDENCE IS HIGH**

- **TECHNOLOGY IS READY FOR SELECTIVE APPLICATION OF COMPOSITES IN NEW AIRCRAFT**

- **COMPONENTS PROGRAM PROVIDED FOCUS FOR ADVANCED RESEARCH**

DURABILITY OF AIRCRAFT COMPOSITE MATERIALS

H. Benson Dexter

NASA Langley Research Center
Hampton, Virginia

INTRODUCTION

Since the early 1970's, the NASA Langley Research Center has had programs under way to develop a data base and establish confidence in the long-term durability of advanced composite materials for aircraft structures. A series of flight service programs are obtaining worldwide service experience with secondary and primary composite components installed on commercial and military transport aircraft and helicopters. Included are spoilers, rudders, elevators, ailerons, fairings and wing boxes on transport aircraft and doors, fairings, tail rotors, vertical fins, and horizontal stabilizers on helicopters. Materials included in the evaluation are boron/epoxy, Kevlar/epoxy, graphite/epoxy and boron/aluminum. Inspection, maintenance, and repair results for the components in service are reported. The effects of long-term exposure to laboratory, flight, and outdoor environmental conditions are reported for various composite materials. Included are effects of moisture absorption, ultraviolet radiation, and aircraft fuels and fluids. Figure 1 summarizes some points of the aircraft composite materials program.

DURABILITY OF AIRCRAFT COMPOSITE MATERIALS

- FLIGHT SERVICE OF COMPOSITE COMPONENTS
  - TRANSPORT AIRCRAFT
  - HELICOPTERS

- ENVIRONMENTAL EFFECTS ON COMPOSITES
  - WORLDWIDE GROUND-BASED OUTDOOR EXPOSURE
  - FLIGHT EXPOSURE OF MATERIAL COUPONS
  - CONTROLLED LABORATORY EXPOSURE

Figure 1

FLIGHT SERVICE COMPOSITE COMPONENTS ON TRANSPORT AIRCRAFT

Confidence in the long-term durability of advanced composites is being developed through flight service of numerous composite components on transport aircraft. Emphasis has been on commercial aircraft because of their high utilization rates, exposure to worldwide environmental conditions, and systematic maintenance procedures. The composite components currently being evaluated on transport aircraft are shown in figure 2. Eighteen Kevlar/epoxy fairings have been in service on Lockheed L-1011 aircraft since 1973. In April 1982 eight graphite/epoxy ailerons developed under the NASA ACEE program were installed on four L-1011 aircraft for service evaluation. One hundred and eight graphite/epoxy spoilers have been in service on six different commercial airlines in worldwide service since 1973. Thirteen graphite/epoxy DC-10 upper aft rudders are in service on five commercial airlines and three boron/aluminum aft pylon skins have been in service on DC-10 aircraft since 1975. Ten graphite/epoxy elevators have been in service on B-727 aircraft since 1980. In addition to the commercial aircraft components shown in figure 2, two boron/epoxy reinforced aluminum center-wing boxes have been in service on U.S. Air Force C-130 transport aircraft since 1974.

Figure 2

ACEE COMPOSITE SECONDARY STRUCTURES

The three major U.S. commercial transport manufacturers have been under NASA contract to design, fabricate, and test the major secondary composite components shown in figure 3. The components were developed as part of the NASA Aircraft Energy Efficiency (ACEE) Program. Each of the components has been certified by the FAA and flight service evaluation is under way. The three components utilize different design concepts. The graphite/epoxy rudders are multi-rib stiffened, the elevators are constructed with graphite/epoxy skins and Nomex honeycomb sandwich, and the aileron design features a syntactic-core sandwich with graphite/epoxy facesheets. An overall mass saving of 25 percent was achieved for the three components when compared to the production aluminum designs.

Figure 3

FLIGHT SERVICE COMPOSITE COMPONENTS ON HELICOPTERS

Composite components are being evaluated in service on commercial and military helicopters, as shown in figure 4. Forty shipsets of Kevlar/epoxy doors and fairings and graphite/epoxy vertical fins are being installed on Bell 206L commercial helicopters for 5 to 10 years of service evaluation. The helicopters are operating in diverse environments in Alaska, Canada, and the U.S. Gulf Coast. Selected components will be removed from service for residual strength testing. Ten tail rotors and four horizontal stabilizers will be removed from S-76 production helicopters to determine the effects of realistic operational service environments on composite primary helicopter components. Static and fatigue tests will be conducted on the components removed from service and the results will be compared with baseline certification test results. In addition, several hundred composite coupons exposed to the outdoor environment will be tested for comparison with the component test results. A Kevlar/epoxy cargo ramp skin is being evaluated on a U.S. Marine Corps CH-53D helicopter. The laminated fabric skin may encounter severe handling such as rough runway abrasion and impact. Maintenance characteristics of the Kevlar skin will be compared with those of production aluminum skin.

206L DOORS, FAIRING AND VERTICAL FIN

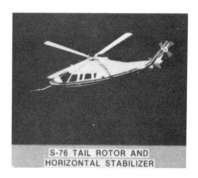
S-76 TAIL ROTOR AND HORIZONTAL STABILIZER

CH-53 CARGO RAMP SKIN

Figure 4

BELL 206L HELICOPTER COMPOSITE COMPONENTS

The four composite components that are being evaluated on the Bell 206L are shown in figure 5. Four different structural design concepts are used in the composite components. The forward fairing is a sandwich structure with a single ply of Kevlar/epoxy fabric co-cured on a polyvinylchloride foam core. The vertical fin is constructed with graphite/epoxy facesheets bonded to a Fibertruss honeycomb core. The litter door is constructed with Kevlar/epoxy fabric and has local reinforcement at load introduction points (hinges and latch assembly). The litter door is a hollow section with inner and outer skins and unidirectional Kevlar/epoxy tape is used in the post area and in the hat-section stiffeners. The baggage door is constructed with Kevlar/epoxy fabric facesheets and Nomex honeycomb core. Additional reinforcements are added in the area of the latch and along the edges.

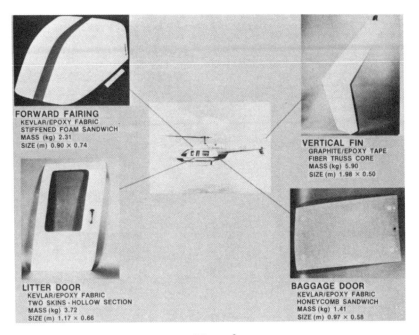

Figure 5

SIKORSKY S-76 HELICOPTER COMPOSITE COMPONENTS

The two composite components that are being evaluated on the Sikorsky S-76 are shown in figure 6. The composite components are baseline designs for the S-76 and are currently in commercial production. The tail rotor has a laminated graphite/epoxy spar with a glass/epoxy skin. The horizontal stabilizer has a Kevlar/epoxy torque tube with graphite/epoxy spar caps, full-depth Nomex honeycomb sandwich core, and Kevlar/epoxy skins. Components have been removed from helicopters and tested after 2 years of service. In addition, small coupons exposed to the outdoor environment were tested and results were compared to baseline values. No significant reduction in either component or coupon strengths was noted. Measured moisture levels were similar to results obtained from other environmental effects programs sponsored by NASA Langley.

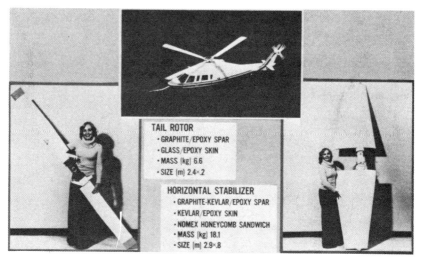

Figure 6

NASA COMPOSITE STRUCTURES FLIGHT SERVICE SUMMARY

A total of 300 composite components have been in service with numerous operators, including foreign and domestic airlines, the U.S. Army, the U.S. Marines, and the U.S. Air Force. The NASA Flight Service Program was initiated in 1973 for the components indicated in figure 7. Over 2.5 million component flight hours have been accumulated with the high-time aircraft having more than 24,000 hours. Some of the graphite/epoxy DC-10 upper aft rudders have been accumulating flight service time at a rate of over 300 hours per month during the past 6 years. The 108 graphite/epoxy spoilers installed on B-737 aircraft have accumulated the highest total component flight hours, over 1.7 million, during 9 years of service. Over 66,000 total component flight hours have been accumulated on the 206L and S-76 composite helicopter components.

AIRCRAFT COMPONENT	TOTAL COMPONENTS	START OF FLIGHT SERVICE	CUMULATIVE FLIGHT HOURS	
			HIGH TIME AIRCRAFT	TOTAL COMPONENT
L-1011 FAIRING PANELS	18	JANUARY 1973	23 130	409 590
737 SPOILER	108	JULY 1973	24 290	1 712 470
C-130 CENTER WING BOX	2	OCTOBER 1974	6 080	12 100
DC-10 AFT PYLON SKIN	3	AUGUST 1975	19 240	55 810
DC-10 UPPER AFT RUDDER	13*	APRIL 1976	22 420	186 970
727 ELEVATOR	10	MARCH 1980	7 180	65 630
L-1011 AILERON	8	MARCH 1982	1 180	7 210
S-76 TAIL ROTORS AND HORIZONTAL STABILIZERS	14	FEBRUARY 1979	3 670	32 070
206L FAIRING, DOORS, AND VERTICAL FIN	124**	MARCH 1981	900	34 000
GRAND TOTAL	300			2 515 850

JUNE 1982

* 7 MORE RUDDERS TO BE INSTALLED
** 36 MORE COMPONENTS TO BE INSTALLED

Figure 7

RESIDUAL STRENGTH OF GRAPHITE/EPOXY SPOILERS

The large number of spoilers with graphite/epoxy skins allows planned retrievals from flight service without seriously impairing the total exposure. Six spoilers, which include two of each of three material systems used in fabricating the spoilers, are selected at random for removal from service annually. The six spoilers are shipped to Boeing for ultrasonic inspection. Three of the spoilers are returned to service after inspection and three are tested to failure to compare residual strengths with the strength of 16 new spoilers that were tested early in the program. Tests have been completed on all three graphite/epoxy systems after 6 years of service and the seventh-year test has been completed on a spoiler constructed with T300/5209. Results of tests conducted to date are shown in figure 8.

The strengths for the individual spoilers generally fall within the same scatter band as was defined by strengths of the new spoilers. The results indicate essentially no degradation in strength after the 7-year period of service for the materials indicated. In addition, stiffness measurements for the graphite/epoxy spoilers indicate essentially no reduction in stiffness as a result of 7 years of service exposure.

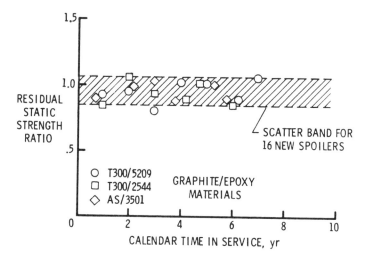

Figure 8

NASA COMPOSITE COMPONENT INSPECTION AND MAINTENANCE RESULTS

The composite components in the NASA Flight Service Evaluation Program are being inspected at periodic intervals to check for damage, defects, or repairs that may occur during normal aircraft operation. The composite components are being inspected by the aircraft operators and manufacturers, and in some cases both visual and ultrasonic inspection methods are being used, as indicated in figure 9.

The Kevlar/epoxy fairings on the L-1011 aircraft have incurred minor impact damage from equipment and foreign objects. Fiber fraying and fastener hole elongations have been noted on all the fairings but no repair has been required. The B-737 graphite/epoxy spoilers have encountered several types of minor damage or defects. Included are spar and doubler corrosion, cuts and dents, and delaminations. Some repairs are being prepared by the airlines but most repairs have been done at Boeing. One of the boron/aluminum aft pylon skins on the DC-10 aircraft was removed from service because of surface corrosion. This corrosion is believed to have been caused by improper surface preparation during panel fabrication.

Minor rib-to-skin disbonds have been detected on two DC-10 rudders. Also, three rudders have encountered minor lightning strikes and one rudder was damaged during ground handling. Minor lightning strikes have also been discovered on two graphite/epoxy elevators. One elevator has been damaged by ground handling. Overall, excellent performance has been achieved with the NASA flight service composite components.

COMPONENT	INSPECTION INTERVAL, months	INSPECTION METHODS	STATUS
L-1011 FAIRING PANELS	12	VISUAL	MINOR IMPACT DAMAGE, FIBER FRAYING AND HOLE ELONGATIONS
737 SPOILER	12	VISUAL ULTRASONIC	INFREQUENT MINOR DAMAGE REPAIRED AT BOEING
DC-10 AFT PYLON SKIN	12	VISUAL	ONE SKIN PANEL REMOVED DUE TO CORROSION
DC-10 UPPER AFT RUDDER	3, 12	VISUAL ULTRASONIC	MINOR RIB-TO-SKIN DISBOND ON TWO RUDDERS; MINOR LIGHTNING STRIKE ON THREE RUDDERS; GROUND HANDLING DAMAGE ON ONE RUDDER
727 ELEVATOR	13	VISUAL	MINOR LIGHTNING STRIKE ON TWO ELEVATORS; GROUND HANDLING DAMAGE ON ONE ELEVATOR

Figure 9

B-737 SPOILER IN-SERVICE DAMAGE AND REPAIR

During the first nine years of flight service there have been 67 instances in which graphite/epoxy spoilers have received damage in service sufficient to require repairs. Typical damage includes graphite/epoxy skin blisters, trailing-edge delamination, miscellaneous cuts and dents, and corrosion of the aluminum spar and doublers, as shown in figure 10. Over one-half of the damage incidents were caused by a design problem wherein actuator rod-end interference caused upper surface skin blisters. The actuator rods have been modified to prevent future damage. Nineteen repairs have been required as a result of corrosion damage to the aluminum spar and aluminum doublers. The corrosion initiates at a spar splice and is probably caused by moisture intrusion through a crack in the sealant material coupled with manufacturing defects in the aluminum surface preparation and corrosion protection schemes. Bondline fatigue in the spar splice area probably contributes to crack initiation and subsequent corrosion. There have been no incidents of galvanic corrosion between the graphite/epoxy skins and the aluminum honeycomb substructure. There have been ten incidents of cuts and dents caused by airline use and four trailing-edge delaminations that were apparently caused by normal aircraft maintenance and moisture intrusion.

Overall, excellent in-service performance has been achieved with the graphite/epoxy spoilers. Several of the airline maintenance executives have expressed the opinion that significantly fewer problems have been experienced with the graphite/epoxy spoilers compared to production aluminum spoilers.

PROBLEM	NUMBER OF INCIDENTS	PERCENT OF TOTAL	CAUSE
BLISTER ABOVE CENTER HINGE FITTING	34	51	DESIGN
SPAR AND DOUBLER CORROSION	19	28	DESIGN/MFG.
MISCELLANEOUS CUTS AND DENTS	10	15	AIRLINE USE
TRAILING-EDGE DELAMINATION	4	6	ENVIRONMENT

Figure 10

DC-10 COMPOSITE RUDDER LIGHTNING DAMAGE

Three of the graphite/epoxy upper aft rudders flying on DC-10 aircraft have sustained minor lightning strikes. The rudder that encountered the most severe strike is shown in figure 11. The damage was localized in an area measuring approximately 1.3 cm by 4.0 cm near the trailing edge of the structural box. The paint layer and four of the outer layers of the graphite/epoxy were removed by the lightning strike. Dry graphite fibers around the edge of the damaged region suggested that the epoxy resin had been vaporized by intense heat generated by the lightning strike. Repair of the rudder was performed in accordance with repair procedures established at the time the graphite/epoxy rudders were certified by the FAA. The repair consisted of a fiberglass cloth patch and a room temperature curing epoxy adhesive. The other two rudders were repaired in a similar manner using either fiberglass or graphite cloth. All three repairs were performed by airlines maintenance personnel and the aircraft resumed scheduled airline service.

Figure 11

ENVIRONMENTAL EFFECTS ON LONG-TERM DURABILITY OF
COMPOSITE MATERIALS FOR COMMERCIAL AIRCRAFT

Concurrent with the flight evaluation of structural composite components, NASA initiated programs to determine the outdoor ground, flight, and controlled laboratory environmental effects on composite materials, as shown in figure 12. Included are effects of moisture absorption, ultraviolet radiation, aircraft fuels and fluids, and sustained tensile stress. Specimens configured for various mechanical property tests are exposed to real-time environmental conditions on aircraft in scheduled airline service and at various ground stations around the world. Composite specimens are also exposed to controlled laboratory environments including temperature, relative humidity, and simulated ultraviolet radiation. The tests, involving more than 17,000 composite specimens, are scheduled to run for up to 10 years. The results of these tests will be correlated to provide a broad data base for environmental effects on composite materials.

- GROUND EXPOSURE — 10 YEARS AT AIRPORTS
- FLIGHT EXPOSURE — 10 YEARS ON AIRPLANES
- LABORATORY EXPOSURE — CONTROLLED CONDITIONS
- CORRELATE AND ANALYZE RESULTS

Figure 12

Selected Additional Subjects

WORLD-WIDE ENVIRONMENTAL EXPOSURE OF COMPOSITE MATERIALS

Composite test specimens are being exposed to outdoor environmental conditions at the ground station locations shown in figure 13. Specimens are mounted on racks and positioned on building rooftops where they are exposed to ambient environmental conditions. Test specimens are configured for interlaminar shear, flexure, compression, and tension tests. Stressed and unstressed tension specimens are being exposed to assess the effects of sustained tensile load. Some specimens are unpainted to evaluate the effects of weathering on unprotected resin matrix materials, while other specimens are painted to evaluate protection afforded by standard aircraft paint. The materials being evaluated include several different graphite/epoxy and Kevlar/epoxy systems. Specimens are removed from the racks at intervals of 1, 2, 3, 5, 7, and 10 years to evaluate mass and mechanical property changes.

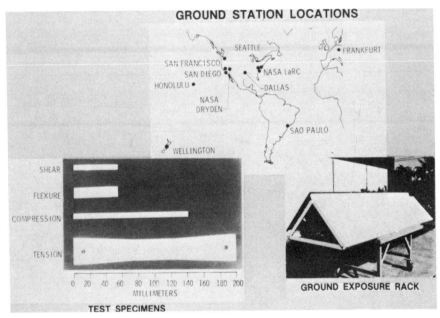

Figure 13

MOISTURE ABSORPTION DURING GROUND EXPOSURE

The moisture contents of four graphite/epoxy and two Kevlar/epoxy material systems after 5 years of exposure at six exposure sites are shown in figure 14. The data shown were obtained from flexure specimens that were exposed on outdoor racks located at Hampton, Virginia; San Diego, California; Sao Paulo, Brazil; Wellington, New Zealand; Honolulu, Hawaii; and Frankfurt, Germany. Each point plotted represents an average value for eighteen specimens, three at each of the six locations. The graphite/ epoxy materials have stabilized after 5 years but the Kevlar/epoxy materials are apparently still gaining a slight amount of moisture. The Kevlar/epoxy materials and T300/2544 have moisture levels of about 2 percent. AS/3501 graphite/epoxy has a moisture content of about 1 percent, while both T300/5209 and T300/5208 graphite/ epoxy have moisture contents of about 0.5 percent. The low value in the 5 year scatter band, in all cases, represents specimens exposed in Frankfurt, Germany; the high value for all material systems except T300/5209 represents specimens exposed in Sao Paulo, Brazil. Additional moisture absorption data will be obtained after 7 and 10 years of outdoor exposure.

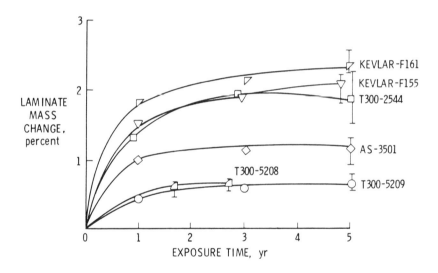

Figure 14

RESIDUAL STRENGTH OF COMPOSITE MATERIALS AFTER
WORLDWIDE OUTDOOR EXPOSURE

Data obtained to date on specimens from four graphite/epoxy and two Kevlar/epoxy systems are shown in figure 15. The data points represent a comparison of the average strength values at six exposure sites with the average baseline strength value for that material system. The shaded area represents a plus-or-minus 10 percent scatter in the baseline strength values. Results of flexure tests indicate little or no degradation in strength over the 7-year exposure period. Compression strengths indicate a slight downward trend, but are still close to the baseline values after 7 years. Short beam shear strength is apparently influenced more by outdoor environmental exposure. The shear strengths for the T300/2544 graphite/epoxy and Kevlar/F-155 systems have dropped below the scatter band of the baseline test results. All the results presented in figure 15 are for unpainted specimens and several of the materials show evidence of surface deterioration due to solar radiation exposure. It is expected that the flexure strength will start to degrade as more matrix resin is leached away and more surface fibers become free. The data obtained to date confirm that the short beam shear strength tests are more sensitive to variations in matrix properties than the flexure or compression tests. One additional set of test specimens remains to be tested after 10 years of outdoor exposure.

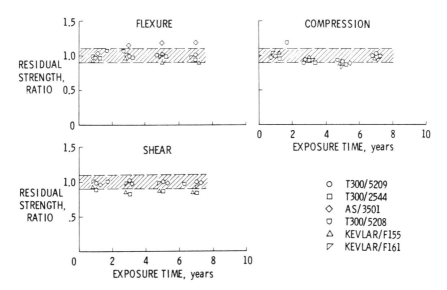

Figure 15

RESIDUAL TENSILE STRENGTH AFTER SUSTAINED STRESS OUTDOOR EXPOSURES

Effects of sustained stress during outdoor environmental exposure are evaluated by exposing tension specimens to 40 percent of ultimate baseline strength. Residual tensile strengths of T300/5208 quasi-isotropic laminated specimens after 7 years of outdoor exposure at the Langley Research Center and San Francisco are shown in figure 16. The residual tensile strength is within the scatter band for the strength of unexposed specimens. Results indicate that the T300/5208 quasi-isotropic tensile specimens were unaffected by either outdoor environment or sustained tensile stress at the two exposure sites indicated. Additional data will be obtained after 10 years of outdoor exposure.

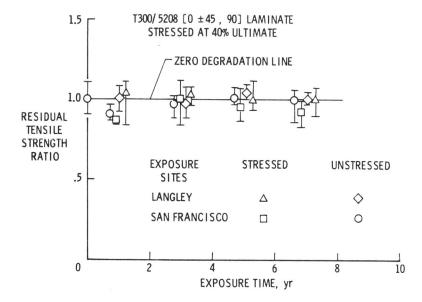

Figure 16

EFFECT OF AIRCRAFT FLUIDS ON COMPOSITE MATERIALS AFTER 5 YEARS OF EXPOSURE

Although aircraft composite structures are exposed almost continuously to various levels of moisture in the atmosphere, they are frequently exposed to various other fluids used in aircraft such as fuel and hydraulic fluid. The effects of various combinations of these fluids on composite materials have been evaluated after 5 years of exposure. Specimens were exposed to six different environmental conditions as follows: ambient air, water, JP-4 fuel, Skydrol, fuel/water mixture, and fuel/air cycling. The water, JP-4 fuel, and Skydrol were replaced monthly to maintain fresh exposure conditions. Specimens exposed in the fuel/water mixture were positioned with the fuel/water interface at the center of the test specimens. The fuel/air cycling environment consisted of 24 hours of fuel immersion followed by 24 hours of exposure to air. Residual tensile strengths of T300/5208 graphite/epoxy, T330/5209 graphite/epoxy, and Kevlar 49/5209 specimens after exposure to the six environments are shown in figure 17. The residual tensile strength of T300/5208 was not degraded by any of the six environments indicated in figure 17. The most degrading environment on the T300/5209 and Kevlar 49/5209 materials was the fuel/water combination. The T300/5209 specimens lost about 11 percent in tensile strength, whereas the Kevlar 49/5209 specimens lost about 25 percent in tensile strength. The ambient air results are consistent with other data obtained from the NASA Langley sponsored ground and flight environmental studies. The tests reported in figure 17 were more severe than actual aircraft flight exposures and the results should represent an upper bound on material property degradation.

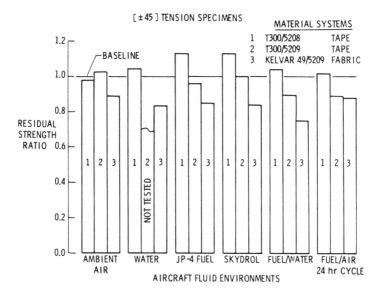

Figure 17

SURFACE DEGRADATION OF AS/3501 GRAPHITE/EPOXY

Scanning electron micrographs were taken of AS/3501 graphite/epoxy flexure specimens with no outdoor exposure and after 5 years of outdoor exposure. The micrograph shown on the left of figure 18 indicates that all the surface fibers are coated with resin for the specimen with no outdoor exposure. The micrograph shown on the right of figure 18 indicates that the surface fibers are no longer coated with resin after 5 years of outdoor exposure. The 5 years of weathering has removed the outer layer of resin and bare graphite fibers are visible.

As with controlled laboratory weatherometer results, these micrographs substantiate the need to keep graphite/epoxy composite aircraft structures painted to prevent ultraviolet radiation damage to composite matrix materials.

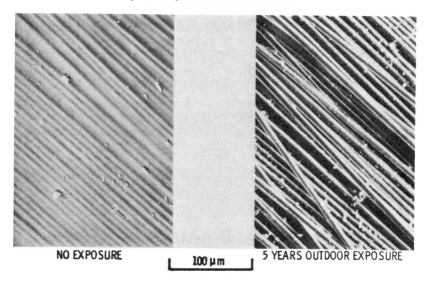

NO EXPOSURE | 100 μm | 5 YEARS OUTDOOR EXPOSURE

Figure 18

CONCLUDING REMARKS

The NASA Langley Research Center has sponsored design, development, and flight service evaluation of 300 composite aircraft components. Excellent in-service performance and maintenance experience have been achieved during 9 years and over 2.5 million hours of flight service. No significant degradation has been observed in residual strength of composite components or environmental exposure specimens after 7 years of service or exposure. Results obtained to date indicate that composite surfaces must be painted with standard aircraft polyurethane paint to protect the matrix from ultraviolet degradation. Test results also indicate that Kevlar/epoxy composites absorb more moisture than most widely used graphite/epoxy composites and a larger reduction in residual strength results for the Kevlar composite systems. Additional details on the programs discussed herein can be found in references 1 through 8.

Confidence developed through NASA-sponsored service evaluation, environmental testing, and advanced composite component development programs has led commercial transport and helicopter manufacturers to make production commitments to selected composite components (fig. 19).

- EXCELLENT IN-SERVICE PERFORMANCE AND MAINTENANCE EXPERIENCE HAVE BEEN ACHIEVED WITH OVER 300 COMPOSITE COMPONENTS DURING 9 YEARS AND OVER 2.5 MILLION HOURS OF FLIGHT SERVICE

- NO SIGNIFICANT DEGRADATION HAS BEEN OBSERVED IN RESIDUAL STRENGTH OF COMPOSITE COMPONENTS OR ENVIRONMENTAL EXPOSURE SPECIMENS AFTER 7 YEARS OF SERVICE OR EXPOSURE

- CONFIDENCE DEVELOPED THROUGH NASA SERVICE EVALUATION, ENVIRONMENTAL TESTING, AND ADVANCED COMPOSITE COMPONENT DEVELOPMENT PROGRAMS HAS LED COMMERCIAL TRANSPORT AND HELICOPTER MANUFACTURERS TO MAKE PRODUCTION COMMITMENTS TO SELECTED COMPOSITE COMPONENTS

Figure 19

REFERENCES

1. Lehman, G. M.: Flight-Service Program for Advanced Composite Rudders on Transport Aircraft. First Annual Summary Report, McDonnell Douglas Corporation, NASA CR-145385, July 1977.

2. Dexter, H. Benson: Composite Components on Commercial Aircraft. NASA TM-80231, March 1980.

3. Dexter, H. Benson and Chapman, Andrew J.: NASA Service Experience with Composite Components. Presented at the 12th National SAMPE Technical Conference, Seattle, WA, October 7-9, 1980.

4. Dexter, H. Benson: Durability of Commercial Aircraft and Helicopter Composite Structures. Proceedings of the Critical Review: Techniques for the Characterization of Composite Materials, AMMRC MS 82-3, May 1982.

5. Kizer, J. A.: Program for Establishing Long-Time Flight Service Performance of Composite Materials in the Center Wing Structure of C-130 Aircraft-Phase V-Flight Service and Inspection, NASA CR-165770, October 1981.

6. Stone, R. H.: Flight Service Evaluation of Kevlar-49/Epoxy Composite Panels in Wide-Bodied Commercial Transport Aircraft-Eighth Annual Flight Service Report, NASA CR-165841, March 1982.

7. Tanimoto, E. Y.: Effects of Long-Term Exposure to Fuels and Fluids on Behavior of Advanced Composite Materials, NASA CR-165763, August 1981.

8. Coggeshall, Randy L.: 737 Graphite Composite Flight Spoiler Flight Service Evaluation, Seventh Annual Report, NASA CR-165826, March 1982.

Other Noyes Publications

ADVANCED CERAMIC MATERIALS
Technological and Economic Assessment

Based on Studies by

Charles River Associates Incorporated
U.S. Department of Commerce International Trade Administration
National Research Council National Materials Advisory Board

This assessment of the current competitive status of advanced ceramic materials presents the situation both from the technological aspect and the economic viewpoint. U.S. prospects relative to Japanese technology and the Japanese governmental philosophy, as well as the general international picture, are outlined and discussed.

Five specific applications of advanced ceramic materials—heat engines, capacitors, integrated optic devices, gas sensors, and cutting tools—are covered at length.

A condensed table of contents listing **part titles and selected chapter titles** is given below.

I. SUMMARY AND CONCLUSIONS
1. Findings Regarding Market Size and Potential Market Growth
2. Findings Regarding Industry Structure
3. Findings Regarding International Differences in Competitive Strategy
4. Potential Economic Benefits of Technological Change in Advanced Ceramic Materials
5. Findings Regarding Technical Barriers

II. HEAT ENGINE APPLICATIONS
6. Structural Ceramic Technology and Technical Barriers
7. Industry and Market Background
8. An Assessment of the Potential Social Benefits of Ceramic Heat Engine Technology

III. CAPACITORS
9. Ceramic Capacitor Technology
10. The U.S. Ceramic Capacitor Industry
11. The Benefits of Future Technological Advance in Multilayer Ceramic Capacitors

IV. INTEGRATED OPTIC DEVICES
12. A Background Discussion of Integrated Optics Technology
13. An Overview of the Integrated Optics Industry
14. The Expected Social Benefits of the Development and Diffusion of Integrated Optical Devices

V. GAS SENSORS
15. An Overview of Ceramic Sensors
16. A Background Discussion of Ceramic Toxic and Combustible Gas Sensor Technology
17. An Overview of the Ceramic Toxic and Combustible Gas Sensor Industry

VI. CUTTING TOOLS
18. A Background Discussion of Technological Issues Concerning Advanced Ceramic Cutting Tools
19. The Advanced Ceramic Cutting Tool Industry
20. An Assessment of the Potential Social Benefits of New Ceramic Cutting Tools

ADDENDUM A. COMPETITIVE ASSESSMENT OF THE U.S. ADVANCED CERAMICS INDUSTRY
21. Trends and Forces Influencing the Future International Competitiveness of the U.S. Advanced Ceramics Industry
22. Foreign Advanced Ceramics Industries and Foreign Governments' Role in Promoting Their Domestic Advanced Ceramics Industries
23. Policy Options to Strengthen the Competitive Position of the U.S. Advanced Engineering Ceramics Industry

ADDENDUM B. COMPETITIVE ASSESSMENT OF THE JAPANESE ADVANCED CERAMICS INDUSTRY
24. Japanese Government Coordination and Management of High-Technology Ceramics Programs
25. Japanese Industrial High-Technology Ceramics Activity

ISBN 0-8155-1037-3 (1985)   651 pages

Other Noyes Publications

CERAMIC MATERIALS FOR ADVANCED HEAT ENGINES
Technical and Economic Evaluation

by

D.C. Larsen, J.W. Adams
IIT Research Institute

J.R. Johnson, A.P.S. Teotia, L.G. Hill
Argonne National Laboratory

Ceramic materials for advanced heat engines are described in this technological and economic evaluation. Advanced power systems for vehicles can potentially produce fuel efficiencies greatly exceeding those of today's gasoline and diesel engines. Current heat engine technologies, however, are limited by problems involving mechanical strength at high temperatures. Structural ceramics, if they can be reliably mass-produced, can make possible improved vehicle fuel efficiencies through higher-temperature operation and reduced vehicle weight.

In Part I of the book state-of-the-art technical ceramics, mainly silicon carbide and silicon nitride, that have potential as structural components in advanced heat engines, are evaluated. Thermal and mechanical property data were generated on candidate materials, and the results were interpreted with respect to microstructure, purity, secondary phases, environmental effects, and processing methods. Properties measured include: flexure strength, elastic modulus, stress-strain, fracture toughness, creep, oxidation, thermal expansion, thermal diffusivity, thermal shock, and stress rupture.

In Part II the macroeconomic impacts of structural ceramics are modeled for two scenarios, one in which the U.S. dominates the commercialization of ceramics in heat engines throughout the 1990s, and the other in which Japan dominates.

A condensed table of contents listing **part titles, chapter titles, and selected subtitles** is given below.

I. TECHNICAL EVALUATION

1. **MATERIALS**
2. **TEST PLAN**
3. **MATERIALS CHARACTERIZATION**
4. **TEST METHODOLOGY**
   Reflected Light Microscopy
   Flexural Strength and Elastic Modulus
   Creep and Stress Rupture
   Dynamic Elastic Moduli
   Thermal Shock/Internal Friction
   Thermal Expansion
   Thermal Diffusivity
5. **MICROSTRUCTURE, ROOM-TEMPERATURE STRENGTH AND ELASTIC PROPERTIES**
   Si_3N_4 Materials
   SiC Materials
6. **ELEVATED TEMPERATURE STRENGTH AND TIME DEPENDENCE**
7. **CREEP BEHAVIOR**
8. **THERMAL EXPANSION**
9. **THERMAL SHOCK RESISTANCE**
10. **OXIDATION BEHAVIOR**
11. **ZIRCONIA CERAMICS FOR DIESEL ENGINES**
12. **FUTURE WORK**

II. ECONOMIC EVALUATION

13. **ECONOMIC IMPACTS OF STRUCTURAL CERAMIC APPLICATIONS**
    Methodology Used to Assess Impacts
    Potential Uses for High-Temperature Structural Ceramics
    Market Penetration for Selected Applications
    Economic Effects of Alternative Scenarios for Commercialization
    Strategic Materials Impacts
    Foreign Competition
14. **THE NEED FOR FEDERAL RESEARCH SUPPORT**
    Assessment of Private Sector Support
    Research and Development: Investment vs. "Borrowing"
15. **INDUSTRY PERSPECTIVES**
    Market Potential for Structural Ceramics
    Need for Research and Development
    Threat of Foreign Competition
16. **CONCLUSIONS**

REFERENCES

APPENDIXES

ISBN 0-8155-1029-2 (1985)   380 pages

Other Noyes Publications

FRACTURE IN CERAMIC MATERIALS
Toughening Mechanisms, Machining Damage, Shock

Edited by
A.G. Evans

Department of Materials Science and Mineral Engineering
University of California, Berkeley

This book presents recent studies on the mechanisms of fracture in ceramic materials—the effects of toughening, machining, and shock. Research on toughening mechanisms, machining and surface damage, thermal shock and general aspects of fracture in ceramic materials is described. Quantitative models of the various fracture processes have been developed. Special emphasis has been placed on the toughening that occurs in the presence of microcracks.

During the last decade, research on the fracture of monolithic single phase and multiphase ceramic polycrystals has attained a maturity which now permits many fracture phenomena to be quantitatively described. Specifically, the predominant fracture initiating flaws have been identified and the fundamental mechanics and statistics related to their fracture severity have been determined. In addition, the crack growth resistance exhibited by common ceramic microstructures can now be expressed in quantitative terms, through the development of micromechanics models of transformation toughening, microcrack toughening, and deflection toughening.

As a result, the next research frontier in the field of advanced monolithic ceramics undoubtedly resides in studies of the processing of optimum microstructures. Progress in this area is summarized in this book. The four main subject areas of the book are toughness/microstructure interactions, machining damage, thermal fracture and reliability, and impact damage.

A condensed table of contents listing **part and chapter titles** is given below.

I. TOUGHNESS/MICROSTRUCTURE INTERACTIONS

1. TOUGHENING MECHANISMS IN ZIRCONIA ALLOYS
2. THE MECHANICAL BEHAVIOR OF ALUMINA: A MODEL ANISOTROPIC BRITTLE SOLID
3. OBSERVATIONS OF INTERGRANULAR, CRACK DEFLECTION TOUGHENING MECHANISMS IN SILICON CARBIDE
4. ON THE CRACK GROWTH RESISTANCE OF MICROCRACKING BRITTLE MATERIALS
5. MICROSTRUCTURAL RESIDUAL STRESSES
6. SPONTANEOUS MICROFRACTURE IN MICROSTRUCTURAL RESIDUAL STRESSES
7. INDUCED MICROCRACKING: EFFECTS OF APPLIED STRESS

II. MACHINING DAMAGE

8. FAILURE FROM SURFACE FLAWS
9. SURFACE FLAWS IN GLASS
10. MECHANISMS OF FAILURE FROM SURFACE FLAWS IN MIXED MODE LOADING
11. GEOMETRICAL EFFECTS IN ELASTIC/PLASTIC INDENTATION
12. RESIDUAL STRESSES IN MACHINED CERAMIC SURFACES
13. FATIGUE STRENGTH OF GLASS: A CONTROLLED FLAW STUDY

III. THERMAL FRACTURE AND RELIABILITY

14. THE THERMAL FRACTURE OF ALUMINA
15. ASPECTS OF THE RELIABILITY OF CERAMICS FOR ENGINE APPLICATIONS

IV. IMPACT DAMAGE

16. LENGTH OF MAXIMAL IMPACT DAMAGE CRACKS AS A FUNCTION OF IMPACT VELOCITY

ISBN 0-8155-1005-5 (1984)   420 pages